Understanding the Digital World

Peter Fettke • Wolfgang Reisig

Understanding the Digital World

Modeling with HERAKLIT

 Springer

Peter Fettke
German Research Center
for Artificial Intelligence (DFKI)
and Saarland University
Saarbrücken, Germany

Wolfgang Reisig
Institut für Informatik
Humboldt-Universität zu Berlin
Berlin, Germany

ISBN 978-3-031-61897-0 ISBN 978-3-031-61898-7 (eBook)
https://doi.org/10.1007/978-3-031-61898-7

This Springer imprint is published by the registered company Springer Nature Switzerland AG
The registered company address is: Gewerbestrasse 11, 6330 Cham, Switzerland

If disposing of this product, please recycle the paper.

Who Builds a House
without Drawing Blueprints?

Leslie Lamport [52]

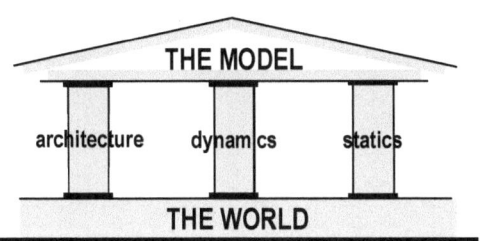

Preface

Several academic disciplines study the *digital* world, including computer science, (business) informatics, (management) information systems, software engineering, and artificial intelligence. In addition to technology-oriented studies, many other aspects of the digital world are researched by human-oriented disciplines, such as arts, humanities, linguistics, medicine, music, and philosophy, to name just a few.

Even a cursory glance at some research streams suggests that there is no common understanding of what constitutes the digital world. One may agree that the digital world is no longer *analog*. But what are its core aspects? *Electronic* data? *Automated* actions? *Virtual* systems? Or something else? It is more than obvious that we lack a common understanding and do not agree on theoretical foundations of the digital world. Moreover, we feel the shortcomings of so far familiar methods. Apparently, new, *digital methods* are required to cope with the challenges of the digital world.

In this book we introduce HERAKLIT – an intuitively straightforward, albeit powerful framework for understanding the digital world. *Architecture*, *dynamics*, and *statics* constitute the three pillars of HERAKLIT. Each concept is introduced explicitly, step by step. Technically, HERAKLIT integrates the proven methods of *algebraic specifications* and *Petri nets* with a new *composition calculus*, resulting in a comprehensive, expressive, concerted, technically simple, digital modeling method. This framework provides the *foundation* for a common understanding of the digital world.

In addition to the conceptual introduction of HERAKLIT, two *case studies* are presented. The first one is about a bakery. This case illustrates all the abstract concepts from a practical perspective. Although all HERAKLIT concepts are illustrated by the first case, the resulting model is not too sophisticated. To demonstrate the usefulness of HERAKLIT for describing the digital world on a large scale, a second case study represents a typical retail business. The intuitive, easy-to-understand structure of HERAKLIT diagrams should not distract from the fact that all HERAKLIT concepts come with a formal, scientific basis.

Saarbrücken, Berlin — *Peter Fettke*
May 2024 — *Wolfgang Reisig*

Acknowledgements

This text passed through a number of iterations, incorporating numerous recommendations of colleagues and students. In particular, we are grateful for remarks and improvements suggested by Dines Bjørner, Dominik Bork, Marc Carwehl, Nachum Dershowitz, Jörg Desel, Andreas Emrich, Hans-Georg Fill, Ulrich Frank, Christoph Freytag, Rob van Glabbeek, Johan Hartung, Holger Hermanns, Stefan Jähnichen, Nijat Mehdiyev, Jan Mendling, Markus Roggenbach, René Rostig, Heinz Schmidt, Elmar Sinz, Mathias Weske, Robert Winter, and many participants of the SUMMER-Soc meetings 2022 and 2023 in Hersonissos, the Petri Net Advanced course 2023 in Toruń, the ER 2023 conference in Hyderabad, the HERAKLIT-Tutorials at the International Conference on Wirtschaftsinformatik 2022 and 2023, the PhD course System Modeling in Informatics and Business Informatics at the Technical University Vienna 2024, and the four FNL-Meetings 2022 and 2023 in Saarbrücken. Finally, we thank for and appreciate the discussions with the members of the Scientific Advisory Board (SAB) of DFKI and the team of Xpect AG, especially Mathias Bauer and Patrick Brandmeier.

We are grateful to the copy editor Philip R. Watson for his many detailed and constructive suggestions. We also thank Ralf Gerstner of Springer-Verlag for his support.

Peter Fettke's work on this book was funded in part by the German Federal Ministry of Education and Research under grant 01IW20006. The authors are responsible for the content of this publication.

Contents

Chapter 1
Introduction

1.1 The digital world

The rise of describing organized handling of items and data

The organized handling of items and data is pivotal for civilizations. This includes administrative rules (Egyptians), mathematical algorithms (*Euklid*), law (Romans), accounting (Italian Renaissance), medicine (clinical pathways), production (*Taylor* principles), traffic rules, military chains of command, and many more. In all these cases, experts demonstrate to scholars how to handle items and data. Historically, there has been little need to describe and to document those procedures in a durable, readable, and explicit form.

The *Industrial Revolution* of the eighteenth century greatly increased the many kinds of human activities and the number of items and data handled. Cooperation between different stakeholders became more important, the amount of (measured) data exploded, and organized activities became more complex. Since then, our civilization's division of labor demanded and produced more precise descriptions of repetitive processes and standardized forms for data and, to a lesser extent, procedures.

We are now in the twenty-first century, in the midst of the *digital* revolution . New principles and methods govern the organized handling of items and data, creating the emerging *digital world*. Of course, these principles and methods require new description techniques for the organized handling of items and data, as well as for the digital world and its components.

The advancing world of computer-integrated systems

Computer-integrated systems meanwhile pervade much of everyday life. Technology for storage, communication, and computation advances quickly, and the *internet of*

P. Fettke, W. Reisig, *Understanding the Digital World*,
https://doi.org/10.1007/978-3-031-61898-7_1

everything becomes real. Meanwhile, computer-integrated systems have become ubiquitous, including (in alphabetic order):

digital avatars, digital banking, digital books, digital communication, digital data, digital education, digital entertainment, digital health, digital identities, digital manufacturing, digital media, digital metering, digital money, digital music, digital phones, digital processes, digital production, digital products, digital publishing, digital retail, digital service machines, digital shops, digital sports, digital systems, digital teaching, digital television, digital tickets, digital twins, digital worlds, et cetera.

In other words: The world has gone digital, and continues to do so.

The digital world represents data in a *digital* form, and describes dynamic change by means of *discrete* states and steps. This allows us to deal with exceedingly large sets of data, with a degree of precision and workability that surpasses analog techniques by magnitudes, and allows for many more applications. It meanwhile also includes areas that typically appear analog, such as music, photography, painting, and film. In addition, modern sensors represent measured values digitally. Business and public administration do likewise with digitally represented information. Terms such as *digitization, digitalization, digital transformation,* and *digital revolution* have arisen, and are frequently used – albeit without too precise a meaning.

Scientific modeling techniques

Classical models in science and in engineering describe dynamic change of real-world systems usually by means of continuous functions over the real number time axis, utilizing the operations of differentiation and integration, and differential equations. Typical examples include the description of acceleration in space, and sensors that represent electrical measurements by means of current or voltage. To underline the importance of these concepts, they are together denoted "The calculus". The calculus is itself already a powerful formal framework. Furthermore, the calculus is the base of numerous techniques and methods for all areas of science and engineering.

As described above, computer-integrated systems are meanwhile predominant. Informatics has delivered a multitude of techniques for modeling such systems. However, consistent, undisputed modeling methods for discrete, digital systems, in analogy to *the calculus* for continuous systems, is missing. This is certainly one of the main reasons for many poor, error-prone, expensive, and even failed projects involving computer-integrated systems. HERAKLIT is intended to contribute to filling this gap.

Modeling in informatics: a historical view

In the 1960s, computers became significantly more powerful. It became increasingly difficult to construct reliable, efficient software for those computers. A *software crisis* was diagnosed [62], and better ways were needed to bridge the gap between the problem to be solved and the software that contributes to solving the problem. A generally accepted therapy was the search for better programming languages, including various bulky, e.g. *Algol 68*, *Ada*, and slim, e.g. *Pascal*, versions. Later on, new programming *paradigms* were proposed such as *functional languages*, e.g. *Haskell*, *logic programming*, e.g. *Prolog*, or *object-oriented languages*, e.g. *Java*. In addition, methodologies and development processes were suggested in the framework of *software engineering*, including *software architectures*, *requirements engineering*, and many more catchwords. Summing up, attempts were made to bridge the gap between real-life problems – e.g. organizing the process of hiring new staff or controlling the machinery of a production line – and software solutions essentially from the perspective of computing and implementable concepts, and not from the perspective of the problems to be solved.

Much less research has started out from the problem-to-be-solved side of the gap. Under the heading of "modeling", methods have been suggested that are problem oriented, yet formal. In order to understand a computer-integrated system, a model of the system should not only cover data and behavior of software, but, most notably, also items, data, and behavior of involved persons, mechanical devices, organizations, et cetera. Furthermore, a system model should describe the *structure* of the system.

To this end, in the 1970s, *algebraic specifications* were suggested; a modeling technique that has its roots in predicate logic. This provided the base for modeling languages such as the *Vienna Development Method* (VDM), later on for the specification language Z, and many others. These languages and methodologies focus on static system aspects, in particular data. They offer only rudimentary support for aspects of behavior and of system architecture.

A different line of modeling techniques concentrates on *behavior*. This includes in particular *Petri nets*, *process algebras*, *statecharts*, and others.

Modeling techniques frequently pay little attention to issues of *system architecture*. Frequently, in graphical representations of models, just a box around some nodes indicates a component.

In the 1990s, the need for a unified conceptualization of the many heterogeneous techniques and languages was manifest, and the *Unified Modeling Language* (UML) [40, 41] was suggested. But UML just collects together more or less informal and poorly integrated concepts.

Since the 1990s, in the field of business informatics many modeling frameworks appeared, e.g. the *Architecture for Integrated Information Systems* (ARIS) [83], *ADONIS* [44], *Business Process Model and Notation* (BPMN) [63], *Event-driven Process Chains* (EPC) [45], *MEMO* [29], the *Semantic Object Model* (SOM) [22], and the *St. Gallen Approach to Business Engineering* [64, 97]. Most of them offer graphical representations to describe behavior, e.g. BPMN, EPC, some provide expressive means for data structures, e.g. ARIS, MEMO, and in some cases also for schemas, e.g. SOM. Some known frameworks use more or less formally defined composition operations. None of the frameworks is fully formally defined, describing many aspects only intuitively. This also applies to frameworks in the information systems discipline, e.g. *Bunge's* ontological model [93] or work system theory [2].

HERAKLIT aims at an integrated framework for all three aspects: systems, actions, and items.

HERAKLIT's view of computer-integrated systems

Computability theory is frequently conceived of as *the* theoretical basis of informatics. Nevertheless, numerous proposals transcend computability theory. A classical reference is [94]. This paper, as well as [4] and [53], propose to transcend computability theory by emphasizing *interaction* as a fundamental computing feature, and asking for a most abstract model of interacting computers. Yet, most of these approaches conceive computer embedded discrete systems from the perspective of computers and software. HERAKLIT agrees on conceiving interaction as a fundamental concept. Additionally, HERAKLIT is intended to contribute a general, formal background to view computer-integrated systems from the perspective of the real world.

1.2 About this book

HERAKLIT combines a research program and a development project, aiming at an infrastructure to model, communicate, and analyze computer-integrated systems [24, 25]. Compared with other frameworks with similar aims, HERAKLIT offers decisive advantages and differences.

HERAKLIT views systems from the perspective of the real world, not from the perspective of computers and software. In particular, a HERAKLIT-modeled system may also include parts that are not intended to be implemented in software, such as person-bounded business processes, or production steps executed by mechanical machines.

HERAKLIT focusses on three aspects of computer-integrated systems:

- HERAKLIT supports the modeling of *large systems*. HERAKLIT offers a general notion of *modules*, together with general concepts for composition and refinement. To cope adequately with large systems, HERAKLIT naturally limits the scope of meaning of any of its symbolic representations to single modules. Thereby, the size of a module can be freely chosen; a module may be large and consist of many smaller modules.
- HERAKLIT assumes *locally bounded* states and events. Consequently, HERAKLIT explicitly models causal dependence and independence of occurrences of events.
- HERAKLIT treats *items and data* of the real or an imagined world, as well as operations on those items and data, on an equal footing. Consequently, symbolic representations of items and of operations are intertwined. Furthermore, the symbolic setting of HERAKLIT is based on concepts of predicate logic.

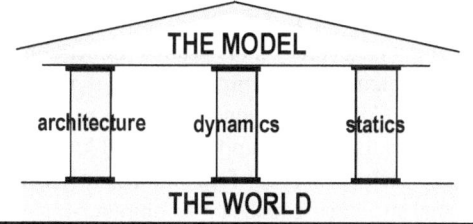

Fig. 1.1 The three pillars of HERAKLIT.

In summary, HERAKLIT integrates these three aspects into three conceptual pillars:

- *Architecture*: architecture modules which can be composed and refined; composition matters!
- *Dynamics*: networked actions, without centralized control or global states; causality matters!
- *Statics*: covering real- or imagined-world items as well as data alike; objects matter!

HERAKLIT starts out from assumptions about discrete systems in the digital world, not from computers, software, et cetera. Nevertheless, HERAKLIT is based on strong theoretical concepts, organized into three pillars: the *composition calculus*, with its fundamental, universal, composition operator, *Petri nets*, with runs that are ordered by causality, and *structures* as used in predicate logic. Figure 1.1 visualizes how the three pillars of HERAKLIT model the real world.

We encourage the readers to visit the HERAKLIT homepage

HERAKLIT.org

any time during their studies of this book. There, they will find more on the conceptual background, more detailed case studies, and further introductory and motivating texts. HERAKLIT concepts are intuitive enough that the case studies given there may be grasped without much formal background.

HERAKLIT is intended to provide a general, formal background to view discrete systems from the perspective of the real world, in analogy to computability theory, which provides a formal background to view discrete systems from the perspective of computers and software.

About HERAKLIT

This book presents the concepts of the HERAKLIT infrastructure. It will enable the reader to understand and design HERAKLIT models. In particular, the reader will be able to model and to integrate aspects of architecture, dynamics, and statics of the real or an imagined world.

This book is easily accessible. It does not require anything beyond an intuitive understanding of the stepwise evolution of systems. Education in classical computer science curricula, emphasizing programming skills, is not too helpful. Likewise, conventional "Theoretical Computer Science", with its symbol-processing automata and functions over integers, is not too supportive. And known models from the area of business informatics lack theoretical foundations. Instead, HERAKLIT spans the range from first informal structuring ideas for a computer-integrated system, over the specification of (business) processes, the contributions of persons, organizations, and mechanical devices, up to the construction of software. The intuitive, easily grasped structure of HERAKLIT diagrams should not distract from the fact that all HERAKLIT concepts come with a formal, well-established scientific basis.

In addition to beginners in the area of system modeling, we expect also readers with previous knowledge. They may get by with studying the Figures, including their accompanying texts.

How to read this book

This book is built up following the three HERAKLIT pillars, starting in Part I with the central HERAKLIT concept of *modules*, in particular their *composition* and *refinement*. Part II covers the second HERAKLIT pillar, concerning *dynamics*, and concentrating on modules that describe aspects of behavior. Part III focuses on *static* aspects. In particular, real- and imagined-world items and their symbolic representation are carefully distinguished and related. Together, the three pillars are consolidated in Part IV, integrating all concepts into a powerful formal framework. The book ends in Part V with a more comprehensive case study, recommendations on how to start modeling with HERAKLIT, and with useful graphical conventions for the graphical representation of HERAKLIT models.

HERAKLIT suggests intuitive, broadly graphical, concepts to represent system architecture, dynamic change, as well as real- and imagined-world items and data. These concepts come with a formal basis; however, HERAKLIT models can be conceived also without detailed knowledge of this formal background. At first reading, the figures with their short texts may be intuitive enough to allow the reader to grasp the essentials.

1.3 Running case study: a bakery

In this volume, all HERAKLIT concepts are motivated and illustrated by the example of modeling various aspects of a bakery business. Considering the business from the angle of pastries produced, the business proceeds in four steps: each pastry is *baked, supplied to the aide, moved to the shop*, and *sold*. The business employs three persons, the *baker*, who bakes pastries, the *aide*, who transports the pastries to the shop, and the *vendor*, who sells the pastries to clients. In a number of variants, we consider different versions of the bakery. The bakery may be refined in different ways, emphasizing different aspects, e.g. the production line of pastries, or the staff. Pastries may vary: We consider the case of three pastries, *bread, cake, pie*, the case of two additional pastries, *roll* and *biscuit*, and the abstract case of refraining from any concrete set of pastries, just assuming the *existence* of a set of pastries. Pastries may be sold in single pieces or in packages.

Most variants pertain to aspects of behavior in the context of many pastries being produced:

- the baker may start baking the second pastry immediately after supplying the first pastry to the aide; more generally, many pastries may be in the process of production at the same time;
- more than one vendor may compete to move a fresh pastry to one of the shops;
- the baker follows a predefined schedule for the order of types of pastries to be baked, or he decides by himself what pastry to bake next;
- the pastries may be supplied with price tags;
- a system is not necessarily commenced, but may be described while running;
- pastries are packed into bags, and bags are sold at fixed prices;
- instead of single pastries, the baker may supply to the aide trays with many pastries.

All these variants may be modeled for fixed sets of pastry and of staff, or for the case of just assuming the existence of pastries and staff.

Part I
The architecture pillar

Each non-trivial computer-integrated system exhibits an *architecture*, i.e. it is composed of sub-systems, sometimes called *parts*, *components*, *constituents*, or similar. Hence, a model of a big system must be composed of sub-models. HERAKLIT covers this fundamental aspect with the notion of *modules*. Modules texture each big system, modules support composition and refinement of sub-modules; modules also structure the behavior of a system.

We start this Part I with the motivation and definition of modules. A detailed representation of the bakery case study shows that, in general, there is no unique or "best" decomposition of a system into modules. Rather, different decompositions emphasize different aspects of the same system. This is followed by a detailed definition of modules and their composition.

Chapter 2
What are modules?
When do we separate them out?
How are they composed?

We compile a list of useful demands on a meaningful concept of "modules" and their composition. This is followed by an example of two different refinements (decompositions) of a bakery. Here, the modules remain abstract. Later on, in Part II, we consider modules that describe specific behavior.

2.1 Requirements for a generic notion of *modules*

HERAKLIT's approach to the notion of modules is most liberal: any set of "elements" and any kind of relations among those elements may constitute a module. Most important is the *composition* of modules; composed modules constitute again a module. A generic notion of composition should likewise be maximally "liberal", i.e. cover each conceivable concrete composition operator as a special case. Altogether, composition of modules should meet the following challenges:

- Composition should be technically simple, but still be capable of expressing any aspect considered relevant when modules are composed.
- Any two modules M and N can be composed, at least technically; though the composed module $M \bullet N$ may not always be useful.
- The composition of modules is frequently accompanied by additional requirements, depending on the actual states or properties of the involved modules; composition may capture alternative behavior, et cetera. All this must be adequately captured by a decent calculus of composition.
- Any network of modules $M_1, \ldots, M_n$ may just be written $M_1 \bullet M_2 \bullet \ldots \bullet M_n$, i.e. brackets must not influence composition.
- Every aspect that is relevant for the effect of composition of modules M and N must be visible in the interfaces of M and N.

HERAKLIT in fact meets all these requirements.

© The Author(s), under exclusive license to Springer Nature Switzerland AG 2024
P. Fettke, W. Reisig, *Understanding the Digital World*,
https://doi.org/10.1007/978-3-031-61898-7_2

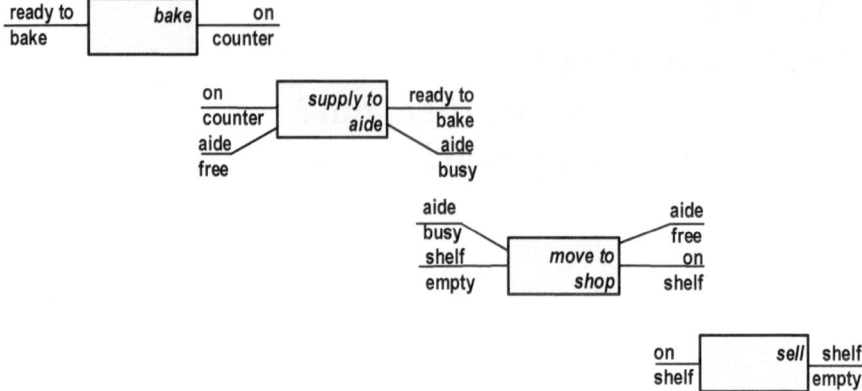

Fig. 2.1 Four modules. Each box represents a *module* and is inscribed with the module's name. Each arc, together with its label, represents a *gate*. The gates, drawn at the left and right margin of the module, constitute the *left* and *right* interface of the module.

2.2 The bakery with four activities

We assume the bakery system to include four essential activities: pastries are *baked*, *supplied* to the aide, *moved* to the shop, and *sold* to clients. Figure 2.1 shows the corresponding four modules in a most abstract setting: each module M is represented by a box, inscribed with the module's name. Some elements of M, called *gates*, are drawn as labeled lines. Touching the box at its left or right margin, a gate belongs to the *left* or *right interface* of the module. Each gate is labeled by a *proposition*. Intuitively, whenever the propositions of a module's left interface gates are met, the corresponding activity occurs and results in the propositions of the right interface; details will follow in Section 5.2, page 54.

A central concern of modules is their *composition*. For intuitively obvious reasons, we want to compose *bake* and *supply to aide*, as in Figure 2.2: the gates labeled *on counter* of *bake* and *supply to aide* are merged. Notice that both modules have a gate labeled *ready to bake*. These gates, however, must not merge!

As already described in the previous Section 2.1, composed modules M and N form again a module, $M \bullet N$. In the module of the composition of *bake* and *supply to aide*, the gate labeled *on counter* goes into the interior of this module; the other gates go to the corresponding interface. Figure 2.3 shows this module.

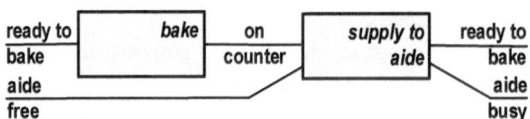

Fig. 2.2 Merging equally labeled gates. The gates with label *on counter* of the modules *bake* and *supply to aide* are merged.

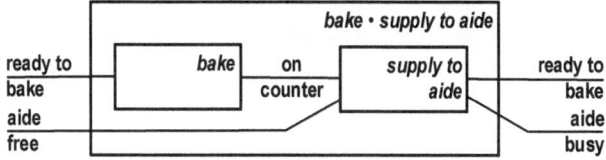

Fig. 2.3 The module *bake • supply to aide*. The gate with label *aide free* in the left interface of the module *supply to aide* has no "partner" gate of the module *bake*. So, this gate forms part of the left interface of the composed module.

Composition of modules is achieved by a general, simple idea: Each module *M* has a *left* and a *right interface*, $^*M$ and M^*. Two modules *M* and *N* are composed by merging equally labeled gates of M^* and $^*N$. Together with the gates of $^*M$, the remaining gates of $^*N$ form the left interface $^*(M • N)$ of *M • N*; and together with the gates of N^*, the remaining gates of M^* form the right interface $(M • N)^*$ of the module *M • N*.

Hence, in Figure 2.3, the left interface $^*(bake • supply to aide)$ includes gates labeled *ready to bake* and *aide free*. And $(bake • supply to aide)^*$ includes gates labeled *ready to bake* and *aide busy*. Likewise, Figure 2.4 shows the composition of three and of all four activities of Figure 2.1, page 14.

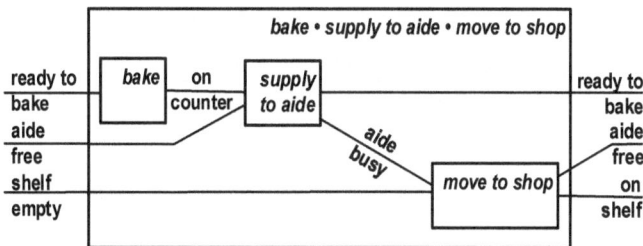

(a) Composition of the three modules yields the module
bake • supply to aide • move to shop.

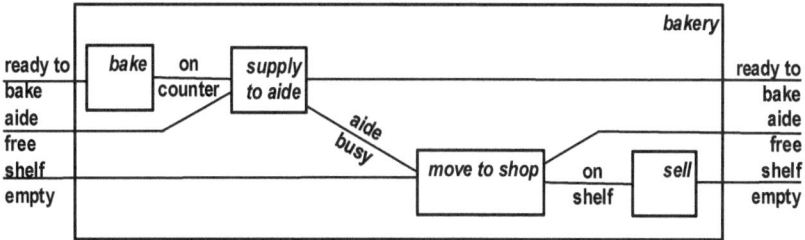

(b) Composition of the four modules yields the module
bake • supply to aide • move to shop • sell.

Fig. 2.4 Two examples for composition.

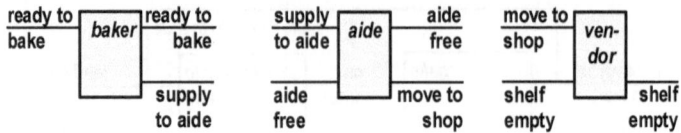

Fig. 2.5 A different view of the bakery: the three staff *baker, aide*, and *vendor*.

2.3 The bakery with three staff

Here we switch focus to a different view of the bakery: now, not four activities, but three staff generate modules: the *baker*, who bakes pastries, the *aide*, who transports the pastries to the shop, and the *vendor*, who sells the pastries to clients. Figure 2.5 shows the corresponding modules.

Content-wise, each gate represents either a *proposition* or an *activity* which the corresponding staff member is engaged in. The three staff will later be refined, representing the respective operational behavior. Intuitively formulated, the behavior of each staff member proceeds from left to right, as follows:

- With the proposition *ready to bake* fulfilled, the baker will bake a pastry and *supply* it to the *aide*; afterwards, the baker is again *ready to bake*.
- The baker shares the activity *supply to aide* with the aide. This activity will occur if the *aide is free*. The aide will move the pastry to the vendor and becomes free for the next pastry.
- The aide shares the activity *move to shop* with the vendor. This activity will occur if the *shelf is empty*. The vendor will sell the pastry, and the shelf becomes empty again.

Figure 2.6 shows examples: Figure 2.6a shows that both the baker and the aide are involved in supplying the pastry to the aide. Figure 2.6b shows that the aide and the vendor are both involved in moving the pastry to the shop. Hence, the bakery can be written as

$$bakery =_{\text{def}} baker \bullet aide \bullet vendor, \tag{2.1}$$

as Figure 2.6c shows. It is interesting to see that the activities are merged: each activity is performed by two staff members.

So far, we considered abstract modules and their composition. Later on, we will describe the operational behavior of the modules. Then it will turn out that the models in Figures 2.4 and 2.6, pages 15 and 17, are different abstract versions of the same operational behavior.

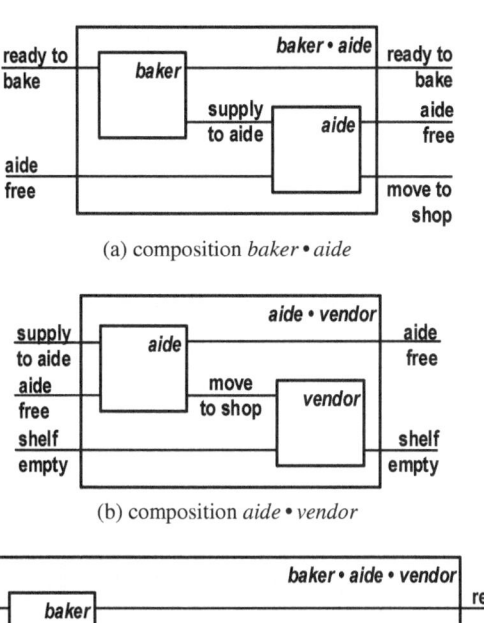

(a) composition *baker • aide*

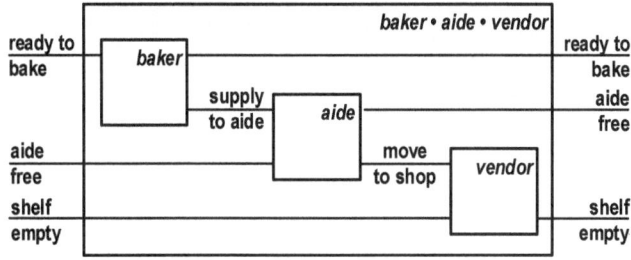

(b) composition *aide • vendor*

(c) composition *baker • aide • vendor*

Fig. 2.6 Composing the modules of staff.

Chapter 3
Fundamental: the notion of *modules*

Conceptually, any *module* can be abstracted to a *graph*, i.e. a module consists of vertices and edges. Each vertex either is an *inner* vertex, or belongs to one of the module's *interfaces*. The examples of the previous Chapter 2 as well as many forthcoming examples and prior experience with HERAKLIT models show that it is quite natural to distinguish a module's *left* and *right* interface. Each vertex in the interfaces is called a *gate* of the module and carries a *label*. Different gates of an interface may carry the same label – although this case does not occur in the examples of the previous Chapter 2. Any graph with two interfaces can be conceived as a module; in extreme cases, an interface is the empty set or a module has no inner vertices.

The well-established pictorial representation of graphs depicts each vertex as a dot, or a circle or a box; an edge links two vertices and is drawn as a line or an arrow between the vertices. For modules, also other graphical representations turn out useful, mainly as shorthand representations. This applies in particular to the Figures in Chapter 2, and will be discussed later.

In this section, we present the general framework of modules and their composition. In order to remain self-contained, we start in Section 3.1 with elementary technical notions of *labeled* and *ordered sets*, thus defining the notion of *interface*. This is followed by *graphs* and their *composition along interfaces* in Section 3.2. Composition is based on the *matching* of labels: equally labeled vertices are merged. Section 3.3 presents the case of modules and their composition. Some fundamental properties of the composition operator, in particular its associativity, are given in Section 3.4. Finally, Section 3.5 presents graphical means for intuitive representations of modules in several contexts, in particular in the context of mixed hierarchies of modules.

P. Fettke, W. Reisig, *Understanding the Digital World*,
https://doi.org/10.1007/978-3-031-61898-7_3

Fig. 3.1 Gates of an interface
are totally ordered. $I =$
$\{a, b, c\}$ and $J = \{d, e, f\}$
are ordered $a < b < c$ and
$d < e < f$. Gate f is labeled
"β"; all other gates are labeled
"α". I and J generate two
matches: $\{a, d\}$ and $\{b, e\}$.
The two remaining gates c
and f do not match.

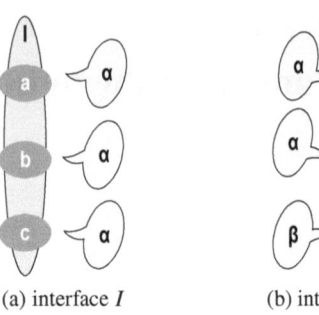

(a) interface I (b) interface J

3.1 Interfaces

An interfaces of a module is a set of module elements called *gates*. A gate is intended
to be merged with gates of other modules. More precisely, gates are merged if they
match, i.e. if they carry the same label and – in case of many likewise labeled gates
– they respect the order of likewise labeled gates.

To characterize interfaces, we assume the common concept of sets, and extend it
by *labels* and *order*:

Definition 3.1 (labeled set) Let A be a set, let Λ be a set with elements called *labels*,
and let l be a mapping assigning each element a of A a label $l(a)$. Then A is *labeled
over* Λ. The mapping l is a *labeling* of A.

An interface will be totally ordered:

Definition 3.2 (order) Let A be a set. A relation $<$ on A is a *strict, total order on*
A, if for all $a, b, c \in A$ holds:

- if $a < b$ and $b < c$ then $a < c$,
- not $a < a$,
- for all $a, b \in A$, it holds that either $a < b$ or $b < a$ or $a = b$.

We usually skip the adjectives "strict" and "total". In most contexts, order describes a "before – after"-relation.

Definition 3.3 (interface) A labeled and ordered set is an *interface*.

With respect to the intended use of interfaces, their elements are called *gates*. In
graphical representations, the gates of an interface are vertically ordered, with the
first, smallest element on top. Figure 3.1 shows two interfaces I and J.

Definition 3.4 (equivalent interfaces) Two interfaces are *equivalent* if and only if
their carrier sets are identical and their orders coincide for equally labeled gates.

Figure 3.2 shows an example of equivalent interfaces. Gates of two interfaces
match, if they coincide in labeling *and* order.

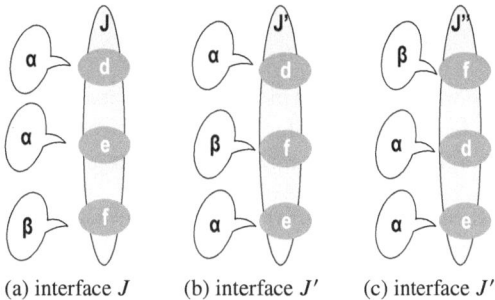

Fig. 3.2 Three equivalent interfaces. Their gates and labelings are identical, and the two "α"-labeled gates d and e are ordered alike. Only the "β"-labeled gate f is ordered differently.

(a) interface J (b) interface J' (c) interface J''

Definition 3.5 (match) Let A and B be two interfaces, let $a \in A$ and $b \in B$. Then $\{a, b\}$ is a *match of A and B*, if for some label λ, both a and b are λ-labeled, and the number of λ-labeled gates that are smaller than a in A is equal to the number of λ-labeled gates that are smaller than b in B. A gate of A that does not belong to a match of A and B is *match free* with respect to B. We use the following notations. Let A and B be two interfaces:

- Let *matches(A, B)* be the set of all matches of A and B.
- Let *matchfree(A, B)* be the set of all gates of A that do not belong to a match of A and B.

Notice that each gate of A belongs either to a match of A and B, or is match free with respect to B.

Figure 3.1 shows the case of two interfaces I and J with two matches. The gate c in I is match free with respect to J and the gate f in J is match free with respect to J:

$$\text{matches}(I, J) = \{\{a, d\}, \{b, e\}\},$$
$$\text{matchfree}(I, J) = \{c\}, \tag{3.1}$$
$$\text{matchfree}(J, I) = \{f\}.$$

It is interesting to see that each of the three equivalent interfaces of Figure 3.2 generates the same matches with the interface I. In fact, in general, it holds:

Theorem 3.1 *Let L, M and N be interfaces, let L and M be equivalent. Then* matches(L, N) = matches(M, N).

Proof. This theorem follows directly from the definition of interface. $\square$

This Theorem offers a degree of freedom for the graphical representation of interfaces.

3.2 Graphs

The well-established pictorial representation of graphs depicts each vertex as a dot, a circle, or a box; an edge links two vertices, and is drawn as a line or an edge between the vertices. We assume the usual notion of directed graphs:

Definition 3.6 (graph) Let V be a set with elements called *vertices*.

1. With two vertices $m, n \in V$, the tuple (m, n) is an *edge over V*.
2. Let E be a set of edges over V. Then $G = (V, E)$ is a *directed graph*.

When obvious from context, the specification "over V" and the adjective "*directed*" are skipped. Most of the following considerations anyway apply also to undirected graphs. In technical terms, we consider an undirected edge as the special case of edges in both directions.

Figure 3.3a, page 23, depicts two graphs, G and H, as usual. Figure 3.3b equips them with the interfaces I and J.

Modules are represented as graphs; hence, the composition of modules is represented as a composition of graphs. To compose two graphs, one firstly selects for each graph an interface. Then for each match, the two vertices of the match are replaced by the match itself. Each edge, so far linked to one of the two vertices, is now linked to the match. Figure 3.3c shows an example.

Definition 3.7 (composition of graphs) Let M and N be two graphs, let $A \subseteq M$ and $B \subseteq N$ be interfaces. Then the *composition of M and N along A and B* is the graph G where:

1. The nodes of G are $(M \setminus A) \cup (N \setminus B) \cup matches(A, B) \cup matchfree(A, B) \cup matchfree(B, A)$.
2. For each edge (x, y) of M or of N,

   - if x and y are both match free, then (x, y) is an edge of G;
   - if x is match free and $\{y, y'\}$ is a match, then $(x, \{y, y'\})$ is an edge of G;
   - if $\{x, x'\}$ is a match and y is match free, then $(\{x, x'\}, y)$ is an edge of G;
   - if $\{x, x'\}$ and $\{y, y'\}$ are matches, then $\{\{x, x'\}, \{y, y'\}\}$ is an edge of G.

Figure 3.3c shows the composition of the two graphs in Figure 3.3b.

Composition of graphs is not associative

At first glance, graphs and their composition might seem to be adequate concepts for modules. However, this would destroy the fundamental requirement of associativity. For example, Figure 3.4, page 23, sketches three graphs L, M and N, with labeled interfaces. The interface of the composition $(L \bullet M) \bullet N$ includes an α-labeled gate that stems from graph N. In contrast, the interface of $L \bullet (M \bullet N)$ includes an α-labeled gate that stems from graph L.

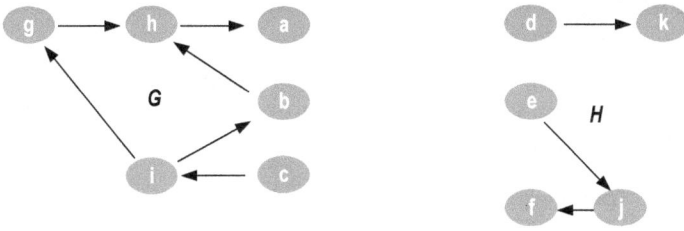

(a) Two graphs G and H.

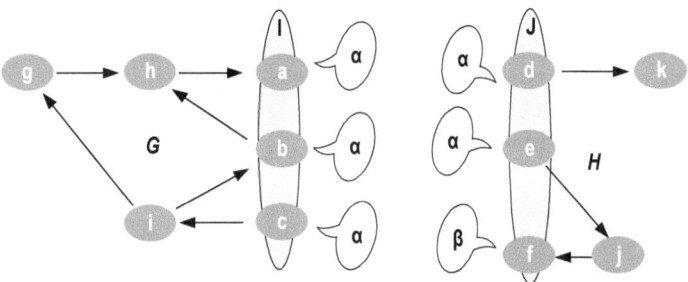

(b) Two graphs with interfaces: graph G with interface I and graph H with interface J.

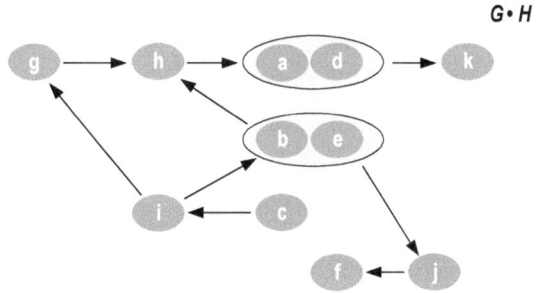

(c) Composition of G and H along the interfaces I and J: the matches $\{a, d\}$ and $\{b, e\}$ become new nodes of $G \cdot H$. The other nodes are not involved in a match and go as nodes to $G \cdot H$.

Fig. 3.3 Graphs G and H with interfaces and their composition.

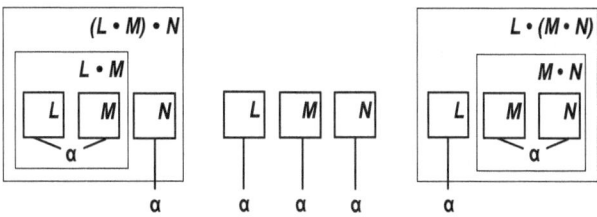

Fig. 3.4 Composition of graphs is *not* associative. Clearly, $(L \cdot M) \cdot N$ and $L \cdot (M \cdot N)$ differ.

3.3 Modules

We are now prepared to define modules and their composition. As discussed in the introductory text of this Section 3, it is useful to distinguish two parts of a module's interface. This leads to the following definition:

Definition 3.8 (module) A *module M* is a graph, together with two interfaces $^*M$ and M^* of nodes of M. The interfaces $^*M$ and M^* are the *left* and *right interface* of M. The nodes $M \setminus (^*M \cup M^*)$ are the *interior* of M.

Graphically, a module is depicted as a box, surrounding the module's graph. The left and the right interfaces are drawn onto the left and right margin of the box. Figure 3.5a, page 25, shows examples.

The above definition does not exclude a gate g of a module to belong both to its left as well as to its right interface.

Definition 3.9 (shared gate) A gate g in $^*M \cap M^*$ is *shared* in M.

In practical contexts, the feature of shared gates is occasionally useful.

A fundamental feature of modules is their composition. We first consider the case of modules without shared gates. To compose two such modules M and N, their underlying graphs are composed along M^* and $^*N$. The interfaces $^*(M \bullet N)$ and $(M \bullet N)^*$ of the composed system $M \bullet N$ expand the interfaces of $^*M$ and N^* by the match free elements of $^*N$ and M^*.

Definition 3.10 (composition of modules without shared gates) Let M and N be two modules.

- The graph of their composition $M \bullet N$ is defined as the composition of the graphs of M and N along the interfaces M^* and $^*N$.
- The left interface $^*(M \bullet N)$ of $M \bullet N$ is $^*M \cup matchfree(^*N, M^*)$. The elements of $^*M$ are ordered before the elements of $matchfree(^*N, M^*)$.
- The right interface $(M \bullet N)^*$ of $M \bullet N$ is $N^* \cup matchfree(M^*, ^*N)$. The elements of N^* are ordered before the elements of $matchfree(M^*, ^*N)$.

Figure 3.5b shows an example of composing two modules.

Turning back to shared gates, it is obviously awkward to draw a shared gate g both onto the left as well the right margin of the box of a module M. Instead, it turns out useful to draw g twice, onto the left and the right margin of the box, and to link these drawings by a double line – reminding the equality symbol "=". Figure 3.6a shows examples. Composition of modules with shared gates moves matches to interfaces:

Definition 3.11 (composition of modules with shared gates) Let M and N be two modules over Λ. To define the composed module $M \bullet N$, the above Definition 3.10 is complemented as follows:

- each $x \in {}^*M \cap M^*$ with a match $\{x, y\}$ of M^* and $^*N$ is in $^*(M \bullet N)$ replaced by $\{x, y\}$;

- each $y \in {}^*N \cap N^*$ with a match $\{x, y\}$ of M^* and ${}^*N$ is in $(M \bullet N)^*$ replaced by $\{x, y\}$.

Figure 3.6b, page 26, shows an example.

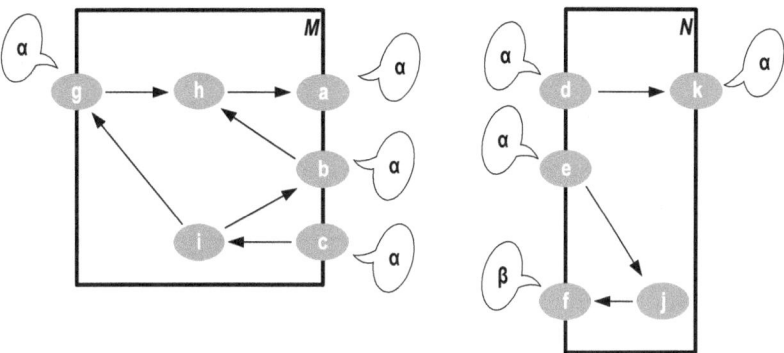

(a) Two modules: Module M with graph G of Figure 3.3, page 23, and interfaces ${}^*M = \{g\}$ and $M^* = \{a, b, c\}$ and module N with graph H of Figure 3.3 and interfaces ${}^*N = \{d, e, f\}$ and $N^* = \{k\}$. M^* and ${}^*N$ are ordered as in Figure 3.1, page 20.

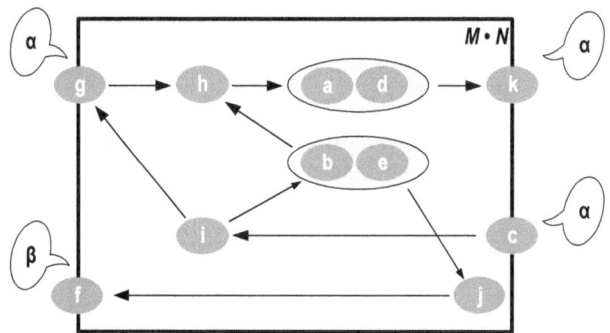

(b) Composition of M and N: $M \bullet N$ with graph $G \bullet H$ of Figure 3.3 and interfaces ${}^*(M \bullet N) = \{g, f\}$ and $(M \bullet N)^* = \{k, c\}$.

Fig. 3.5 Modules M and N and their composition.

The role of labels

Many modeling techniques label interface elements by "complementary" labels. For example, the label "λ", may be complemented by the underlined "$\underline{\lambda}$" or primed "λ'". Composition is then defined by merging complementary labeled interface elements. This reflects the intuition that λ and $\underline{\lambda}$ represent items or properties that complement one another. For example, λ may represent an electric socket, and $\underline{\lambda}$ a plug. HERAKLIT labels occur likewise in two variants, in a module's left or right interface: a gate labeled "λ" in the right interface M^* of a module M may represent an item or property that complements an item represented by a gate labeled "λ" in the left interface $^*N$ of a module N. Returning to the above examples, a gate labeled "λ" in the right interface M^* of a module M may represent a socket; a gate labeled "λ" in the left interface on a module N may represent a plug. Likewise, in M^*: a *positive magnetic pole*, in $^*N$: a *negative pole*; in M^*: a *button*, in $^*N$: a *buttonhole*; in M^*: a *provider*, in $^*N$: a *requester*; in M^*: a *seller*, in $^*N$: a *buyer*; in M^*: a *nub of a Lego brick*, in $^*N$: a *cavity*. Composition along the shared label "λ" can then intuitively be conceived as: to *link electric wires*, to *click magnets together*, to *fasten clothes*, to *transfer items*, to *operate a business*, to *stick Lego bricks together*.

3.4 Properties of composition

The composition operator exhibits several properties that make it practically useful. By far the most important is *associativity*.

Theorem 3.2 *Let L, M, N be three modules. Then* $(L \bullet M) \bullet N = L \bullet (M \bullet N)$.

Proof. This has been proven in [77].   □

Hence, we can write $L \bullet M \bullet N$ without brackets. Without this property, models consisting of many modules would not be practically manageable. Figure 3.7, page 27, shows an example.

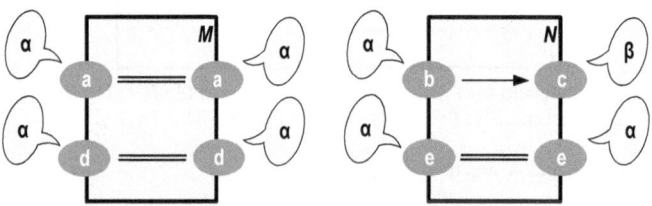

(a) Gates a and d belong to $^*M$ and M^*. Gate e belongs to $^*N$ and N^*.

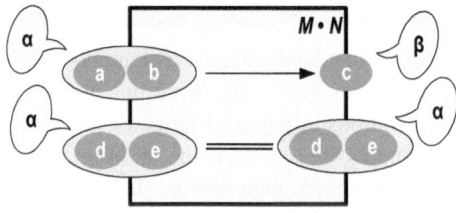

(b) Composition of M and N.

Fig. 3.6 Modules M and N with shared gates and their composition.

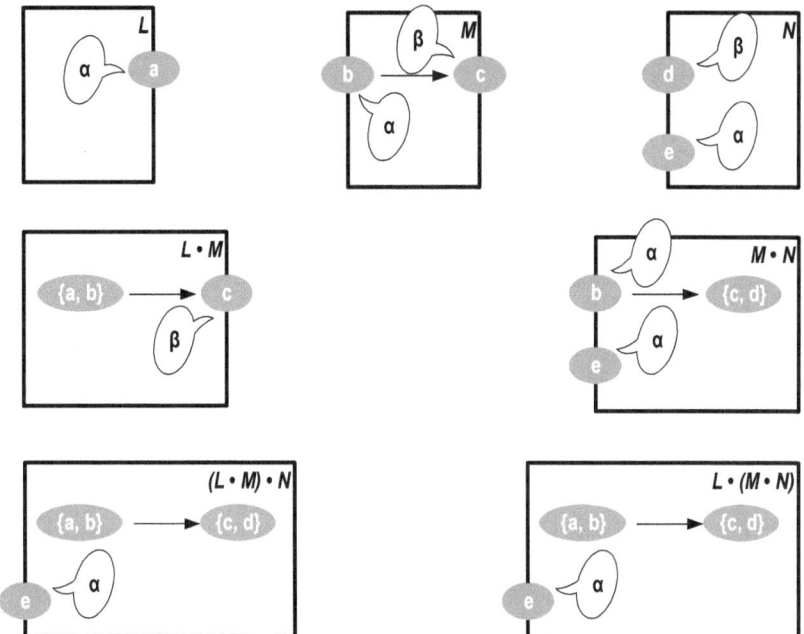

Fig. 3.7 Composition of modules is *associative*: $(L \bullet M) \bullet N = L \bullet (M \bullet N)$.

Composition of two modules M and N without shared gates is commutative *up to the equivalence* if and only if M and N share no equal interface labels:

Theorem 3.3 *For two modules M and N without shared gates holds: $M \bullet N$ and $N \bullet M$ are equivalent if and only if no label occurs in $^*M$ or M^* as well as in $^*N$ or N^*.*

Proof. To prove "→", by contradiction assume equally labeled elements in M^* and in $^*N$. Then there exists a match (m, n) of M^* and $^*N$. Then (m, n) goes to the interior of $M \bullet N$, but not to the interior of $N \bullet M$.

To prove "←", assume no label occurs in $^*M$ or M^*, as well as in $^*N$ or N^*. In this case, $^*(M \bullet N) = {}^*M \cup {}^*N = {}^*N \cup {}^*M = {}^*(N \bullet M)$. Likewise, $(M \bullet N)^* = (N \bullet M)^*$.   □

Figure 3.8, page 28, shows an example.

Strictly, in $M \bullet N$, f is ordered before c. However, as explained in Figure 3.2, page 20, differently labeled gates in an interface may be swapped. We prefer an equivalent, more symmetric representation here.

The rest of this Chapter 3 covers two further aspects of HERAKLIT modules and their composition operator: A graphical convention for "anonymous" vertices, and the special case of modules without inner vertices.

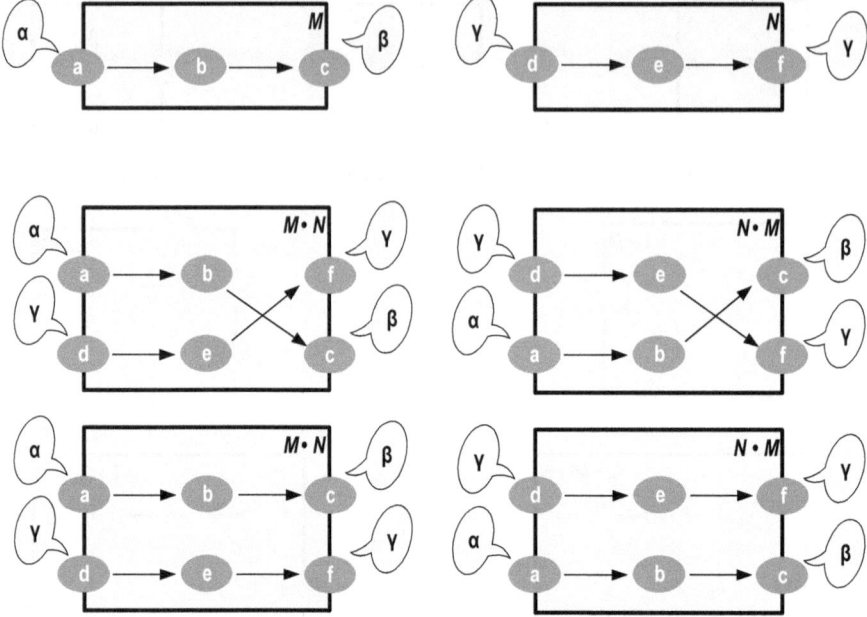

Fig. 3.8 Four equivalent representations of commutative composition of M and N.

Modules spawn monoids

In mathematical terms of universal algebra, a given set Λ defines the set $M(\Lambda)$ of all modules with gate labels taken from Λ. Composition "•" is then a total binary operation $\bullet : M(\Lambda) \times M(\Lambda) \rightarrow M(\Lambda)$, i.e. any two modules in $M(\Lambda)$ can be composed, resulting in a module in $M(\Lambda)$. Furthermore, $M(\Lambda)$ contains a *neutral* module, $M_\emptyset =_{\text{def}} (\emptyset, \emptyset)$, i.e. $M_\emptyset$ contains no vertices and no edges at all. Hence, both interfaces ${}^*M_\emptyset$ and $M_\emptyset^*$ contain no gates. This implies for each module M in $M(\Lambda)$:

$$M \bullet M_\emptyset = M \text{ and } M_\emptyset \bullet M = M. \tag{3.2}$$

Associativity and neutral element together define the algebraic structure of a *monoid*. The well-known monoid of words over an alphabet Λ can be conceived as a special HERAKLIT monoid: To this end, Λ is equipped by a fresh symbol, say ϵ, and each $\lambda \in \Lambda$ is turned into a module with one left and one right gate, both labeled ϵ.

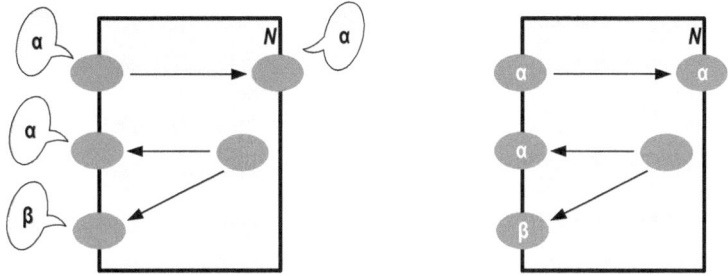

(a) anonymous vertices: identity of a vertex is no longer represented

(b) each gate is inscribed with its label

Fig. 3.9 Module M with vertices rendered anonymous. Here, the individual identity of vertices is no longer visible. Hence, gates can be inscribed with their labels. Context must make it clear whether a gate's inscription denotes its identity or its label.

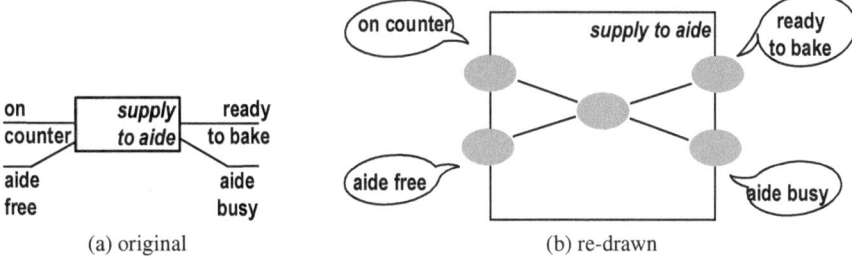

(a) original

(b) re-drawn

Fig. 3.10 Two drawings of the module *supply to aide* of Figure 2.1, page 14.

3.5 Graphical conventions: anonymous vertices

In many models, in particular these in the previous Chapter 2, the identity of gates is irrelevant; what counts are the gates and their labels. Such gates are *anonymous*; in graphical representations, they are usually depicted just as dots as in Figure 3.9a. As an anonymous gate is no longer inscribed with its identity, it can be inscribed with its label instead, as in Figure 3.9b. Context must make it clear whether a gate's inscription denotes its identity or its label.

A further graphical convention has been applied in the representation of the modules in Chapter 2: each gate is anonymous, represented as a line, inscribed with its label. Figure 3.10 depicts this convention. Further conventions will follow later.

3.6 Empty modules

In many contexts it is useful to abstract away from the inner details of a module M, and to stick to the *empty* version $[M]$ of M. The module $[M]$ just skips the interior of M and retains the two interfaces.

Definition 3.12 (empty module, abstraction)

1. A module M is *empty* if and only if it has no inner vertices or edges, i.e. only interfaces with some gates.
2. The *inner abstraction* of a module M is the empty module $[M]$ with interfaces $^*[M] = {}^*M$ and $[M]^* = M^*$.

Interfaces are preserved under any mixture of composition and abstraction.

Theorem 3.4 *Let M and N be modules. Then*

1. $^*([M] \bullet [N]) = {}^*([M] \bullet N) = {}^*(M \bullet N)$;
2. $([M] \bullet [N])^* = (M \bullet [N])^* = (M \bullet N)^*$.

Proof. To prove this Theorem, observe that the interfaces of a composed module depends only on the interfaces of its component modules. Furthermore, the interfaces of a module and of its inner abstraction are identical.   □

We frequently consider empty modules with anonymous gates. Then it is useful to depict the gates not as dots, but as lines. Figures 2.1 and 2.5, pages 14 and 16, in Chapter 2 show examples. This renders composition graphically appealing, as Figures 2.3 and 2.4, pages 15 and 15, show.

The architecture of a system can be built up by composing empty modules. For example, the four activities of Figure 2.1, page 14, thus generate the modules in Figures 2.3, page 15, and Figure 2.4, page 15. As a variant, one also may first assume empty modules *bakehouse* and *salesroom*, and compose them as in Figure 3.11.

Then one may refine the interior of the two modules by

$$bakehouse =_{\text{def}} bake \bullet supply\ to\ aide \text{ and } salesroom =_{\text{def}} move\ to\ shop \bullet sell \quad (3.3)$$

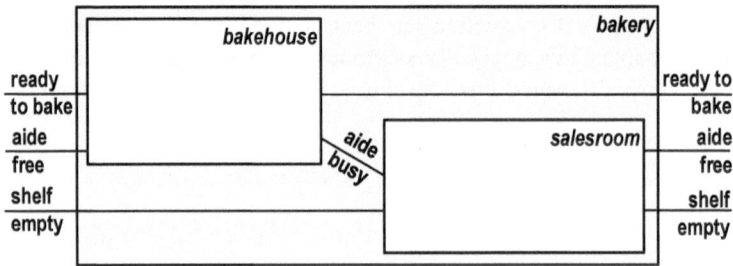

Fig. 3.11 The module *bakery* is composed from the empty modules *bakehouse* and *salesroom*: *bakery* =_{def} *bakehouse* • *salesroom*.

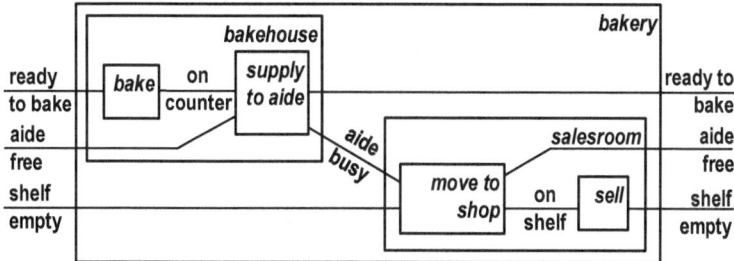

Fig. 3.12 A hierarchy of modules.

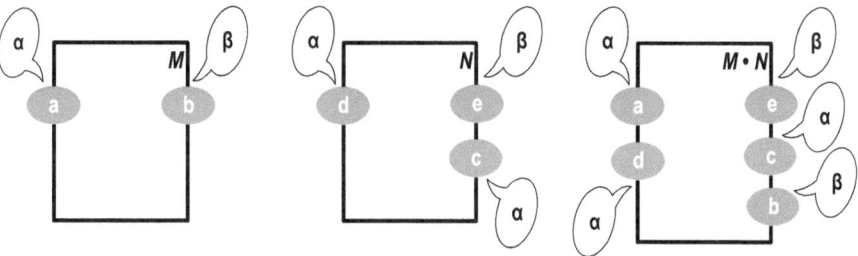

Fig. 3.13 Composition of empty modules yields an empty module if and only if M^* and $^*N$ share no gate labels.

as in Figure 3.12.

The composition $M \bullet N$ of empty modules M and N is empty if no inner elements arise. This holds in the special case of disjoint labels of M^* and $^*N$.

Theorem 3.5 *For empty modules* $[M]$ *and* $[N]$ *holds:* $[M] \bullet [N]$ *is empty if and only if no gates of* M^* *and* $^*N$ *are equally labeled.*

Proof. To prove this theorem, observe that $[M] \bullet [N]$ is empty if and only if there exists no match of $[M]^*$ and $^*[N]$. This holds if and only if there exists no match of M^* and $^*N$. This holds if and only if no gates of M^* and $^*N$ are equally labeled. □

Figure 3.13 sketches this case.

Related work to Part I

The quest for modules and their composition is of paramount importance for any kind of system models. Starting with the famous work of *Parnas* [65], literature on modules and composition of modules is voluminous. In most contributions, models and programs are equipped with interfaces, and composition is defined along those interfaces. Typical examples include:

- *Applied modeling frameworks*: The most influential modeling framework, *UML* [40], suggests, on a semi-formal level, a particular kind of composition: a set of mutually unrelated modules are composed and the result constitutes a larger module. Altogether, this yields a tree-shaped architecture, with vertically unrelated branches.
- *Web services*: There are plenty of approaches to *web service* composition; a composition can be agent-based, synchronous, asynchronous, or based on interaction protocols [72].
- *Interface languages*: An *interface language* represents interface descriptions in a language-independent way, in order to link programs in different languages [51].

More abstract calculi which support modules and their composition include *Hoare's Communicating sequential processes* (*CSP*) [37], and the many versions of process calculi [6]. These calculi come with associative composition, but at the price of non-determinism and will be discussed in the related work of Part II.

Composition of Petri nets has a long tradition, with many contributions. We just mention some relevant ideas of the last three decades: [18] composes many Petri net modules in one go, merging all equally labeled places as well as transitions. Associativity of composition is not addressed; but it seems obvious that this composition operator is not associative.

A long-standing initiative with many variants is algebraic calculi for Petri nets, such as the *box calculus* and *Petri net algebras* in various forms [10]. In the spirit of process algebras, these calculi define classes of nets inductively along various composition operators; merging equally labeled transitions. Associativity of composition is mostly assumed, but rarely explicitly discussed.

[5] defines composition $A \bullet B$ of "composable" nets A and B in a categorical framework, with A and B sharing a common net fragment, C. The effect of composition on distributed runs is studied. The issue of associativity of composition is not addressed. In HERAKLIT, one would formulate the net fragment C as an adapter, constructing $A' \bullet C \bullet B'$. Here, A' and B' are A and B without C.

[48] suggests a general framework for "modular" Petri nets, i.e. nets with features to compose a net with its environment. A module may occur in many instances. An interesting feature is *generators* of data types. Associativity of composition is not discussed in this paper, but implicitly assumed. HERAKLIT is entirely open to specializations of data; so, the idea of generators may be applied in the context of HERAKLIT as well.

Petri nets with two-faced interfaces are discussed in [73, 86]. They suggest *Petri nets with boundaries (PNB)*, with two composition operators. A *PNB* resembles a HERAKLIT module in that it has a left and a right interface. One of the composition operators recalls the HERAKLIT case of transition interfaces with all transitions labeled alike, the other one equals the case of disjoint labels. Interestingly, both operators are associative, but in general not commutative. It might be worthwhile to formulate the paper's results on compositional reachability in the framework of HERAKLIT.

The first ideas for HERAKLIT modules and the HERAKLIT composition operator occurred for runs of Petri nets in [75]. Associativity of the composition operator was proven in [77]. All publications on HERAKLIT, as compiled in [23], employ the composition operator. To the best of our knowledge, the HERAKLIT version with double-faced interfaces and associative composition is unrivaled.

Double-faced interfaces for modeling techniques other than Petri nets can be found in the literature, albeit in versions that are more specialized than the HERAKLIT version. A typical example is the "piping" or "chaining" operator $P >> Q$ for the language CSP [79]. Associativity of $>>$ is assumed without discussion.

There are numerous versions of equipping classical automata with mechanisms for communication; e.g. [19, 21]. Usually, composition includes semantic requirements and properties, yielding lots of variants of compositions $A \bullet B$. In contrast, HERAKLIT comes with a single composition operator, and locates semantic requirements in an adapter, C, in a composition $A \bullet C \bullet B$.

In the context of business informatics and business process management, many ideas to compose system models emerged [84]. Event-driven process chains (EPC), used to describe e.g. the *SAP R/3 system*, suggest *process interfaces* to describe compositions [91]. This idea is implemented in various software tools. For example, the *ARIS Toolset* uses the idea of *definition copies* and *instance copies* to cope with composition. However, a universal and rigorous definition of these concepts is missing.

BPMN introduces a large set of composition concepts, such as *choreography* and *collaboration diagrams*. These ideas are tightly coupled to process models and are not explicitly formally defined. A broader overview of composition by process choreographies is given by [95, p. 259-306]. This work also introduces the concept of a *workflow module* (page 272), which is similar to a HERAKLIT module. However, there are important differences; for example, a workflow module has only one, not two interfaces, and sticks to Petri nets.

In addition, the development of software based on business components is being developed in various lines of research, e.g. [61] published a memorandum for a specification framework for business components. Such a business component has two types of interfaces; one for the services it offers to other components, and one for the services it requires. The framework is not formally described.

Part II
The dynamics pillar

After the architecture pillar of HERAKLIT in Part one, we now turn to the second HERAKLIT pillar, covering aspects of *dynamics*. HERAKLIT bases its description of dynamics on *Petri nets* [76]. Petri nets reappear in HERAKLIT modules that describe dynamics, in various variants. In particular, Chapter 4 discusses occurrences of states and events and their interplay, and Chapter 5 discusses systems that are composed of states and events.

Chapter 4
Runs

The most fundamental notions for dynamics are the *update of states* and the *occurrence of events*. These two notions are tightly entangled: upon occurrence of an event, some actually reached states are *abandoned*, and some actually not reached *states* are *reached*. This section shows the implications of this fundamental idea.

4.1 Pictorial representation of state updates and event occurrences

Following *Petri nets*, we pictorially represent state updates and event occurrences as circles and boxes, and their relationship as arrows. For instance, Figure 4.1a shows the pictorial Petri net representation of an occurrence of the event *bake*: the actually reached state *ready to bake* is abandoned, and the state *on counter* is reached. Figure 4.1b shows correspondingly an occurrence of event *supply to aide*. Here, two states are abandoned, and two states are reached. Figure 4.1c combines the two event occurrences in an intuitively self-evident manner. Notice that the *ready to bake* state is reached at the beginning and, a second time, at the end.

Figure 4.1 shows examples of the Petri net representation of *steps* and *runs*: a step in Figures 4.1a and 4.1b comprises an event occurrence together with the affected state occurrences. A run, Figure 4.1c, composes a set of steps. Later on we will see also other Petri net models, in particular models for systems and system schemata.

The following Section 4.2 presents HERAKLIT modules based on Petri nets. The rest of Chapter 4 studies special such modules, representing the behavioral concepts of *steps* and *runs*.

The principle of locality

It is well known that the size of storage required by a computation of a computable function cannot be estimated in advance. Therefore, to cover

---

all intended computations, the tape of a Turing machine M is conceptually assumed to be *infinite*. Of course, a real-world implementation of M cannot provide such a tape. The best one can do is to expand a given *finite* storage device during computation, whenever *demanded*.

In his seminal PhD thesis [69], *Carl Adam Petri* in 1961 discussed necessary conditions for a modeling framework that covers such unboundedly expandable storage devices. For example, such a device must avoid a centrally generated clock signal, because iterated expansion can eventually cause race conditions. As a further example, a global variable would eventually require additional organizational means to be "accessible everywhere". Formulated positively, any such device eventually must operate in an asynchronous mode. An operational model for asynchronously (co-)operating components must locally constrain causes and effects of events. In particular, the concept of global state changes has no counterpart in asynchronously operating systems. For example, the entire Internet, or any kind of cyber-physical system, cannot be understood as jumping from on global state to a successor state.

These observations motivated the fundamental concepts of Petri nets, with *local* states and *locally confined* transitions. Each HERAKLIT structure is defined in a unique module. This module naturally limits the scope of items, data, and functions of the structure.

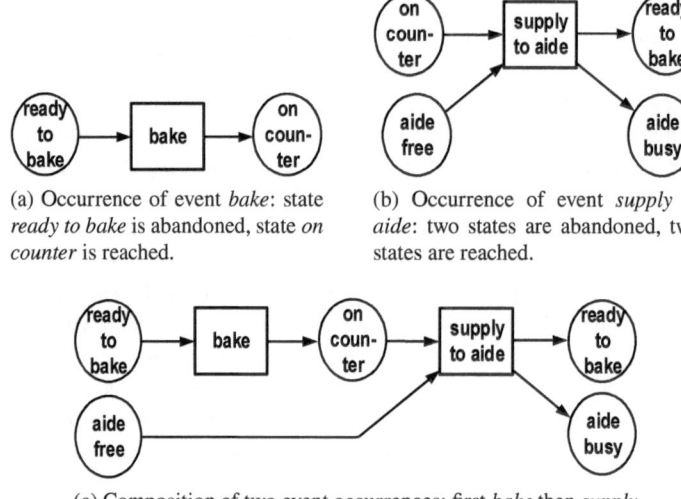

(a) Occurrence of event *bake*: state *ready to bake* is abandoned, state *on counter* is reached.

(b) Occurrence of event *supply to aide*: two states are abandoned, two states are reached.

(c) Composition of two event occurrences: first *bake* then *supply to aide*.

Fig. 4.1 Updates of states caused by occurrence of events.

4.2 Net modules

As discussed in the previous Section 4.1, Petri nets depict passive and active components as circles and boxes, respectively. Arrows represent relations between passive
and active components. This dichotomy governs the formal framework of Petri nets,
and reappears in Petri net-based HERAKLIT modules. All this motivates *net graphs*,
i.e. graphs with two disjoint sets of nodes.

Definition 4.1 (net graph) Let $G = (V, E)$ be a graph, and let P and T be two
disjoint sets with elements called *places* and *transitions*, respectively, such that:

1. $P \cup T = V$;
2. For each edge $(x, y) \in E$ holds: either $x \in P$ and $y \in T$, or $x \in T$ and $y \in P$.

Then G together with P and T is a *net graph*, written $G = (P, T; E)$.

Figure 4.1, page 42, shows examples of very special net graphs. Net graphs are
extended to modules as defined in Section 3.1, page 20.

Definition 4.2 (net module) For a net graph G, a module over G is called a *net
module*.

Hence, a net module is a net graph together with two distinguished interfaces,
i.e. subsets of labeled places and transitions. Furthermore, nodes of an interface are
ordered.

In net modules' interfaces, the label of a place must never coincide with the label
of a transition. This is *generally assumed* in the rest of this monograph: For all
labelings of interfaces of net modules in this monograph, we assume disjoint sets of
labels for places and for transitions.

Theorem 4.1 *The composition $M \bullet N$ of two net modules M and N, is a net module,
too.*

Proof. This theorem holds due to the above general assumption on labelings of
interfaces.   □

4.3 Step modules

As frequently mentioned, we focus on systems which spawn *stepwise* behavior. A *step*
includes the *occurrence of one event*, together with the affected states. HERAKLIT
bonds a step into a special kind of module, called a *step module*. As Figure 4.1,
page 42, already shows, we represent each state occurrence as a place, each event
occurrence as a transition. A step is now represented as a net module M with just
one transition t. The left interface $^*M$ consists of the abandoned states, and the right
interface consists of the newly reached states:

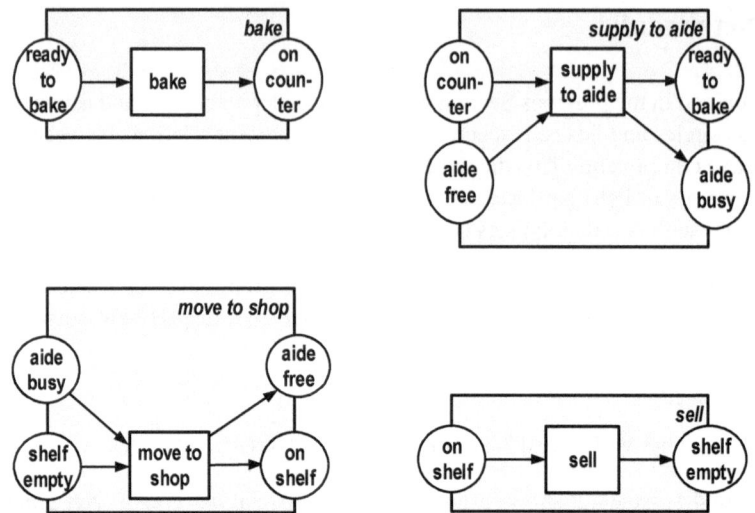

Fig. 4.2 The four step modules of the bakery enterprise. These modules describe the interior of the four activities of Figure 2.1, page 14. Vice versa, the modules of Figure 2.1 are the inner abstractions of the above modules.

Definition 4.3 (step module) Let $M = (P, \{t\}; E)$ be a net module with disjoint interfaces $^*M$ and M^*, such that for each $p \in P$ holds: $(p, t) \in E$ if and only if $p \in {}^*M$, and $(t, p) \in E$ if and only if $p \in M^*$. Then M is a *step module*.

For our running example of a bakery, Figure 4.2 shows four step modules. The step modules *bake* and *supply to aide* turn the net graph of Figures 4.1a and 4.1b, page 42, into step modules. The two step modules *move to shop* and *sell* are now obvious.

Later on, a local state of a module will be understood, in a mathematical framework, as a *proposition*. Occurrence of an event can swap the proposition from "false" to "true" or from "true" to "false".

4.4 Run modules

Composition of steps yields a *run*: intuitively, a run represents evolving behavior by means of composed steps. For example, Figure 4.3 shows the composition

$$bake \bullet supply\ to\ aide \tag{4.1}$$

of the *bake step* module with the *supply to aide* step module of Figure 4.2. In fact, this is Figure 4.1c, page 42, turned into a module. Correspondingly, Figure 4.4 shows the composition

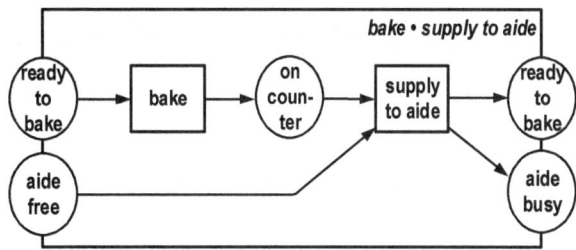

Fig. 4.3 This module composes the first two modules of Figure 4.2, page 44, *bake* and *supply to aide*.

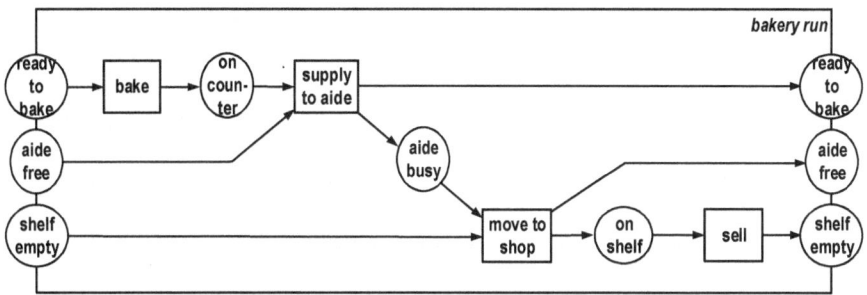

Fig. 4.4 The bakery run, composing all four step modules of Figure 4.2, page 44: *bakery run* =_{def} *bake • supply to aide • move to shop • sell*.

$$bake \bullet supply\ to\ aide \bullet move\ to\ shop \bullet sell \qquad (4.2)$$

of all four step modules of Figure 4.2, page 44. Intuitively, this composed module describes the lifeline of a pastry, from being baked to being sold. Technically, this is reflected by the observation that the places in the left and right interface are labeled alike: The corresponding states are reached again.

Generally, each finite run module can be composed from step modules. However, we will also consider infinite runs. A finite or infinite run module is a special net module. Places and transitions are labeled by states and events, respectively. Equally labeled elements represent different occurrences of the same state or event.

Definition 4.4 (run module) Let $R = (P, T; E)$ be a net module. R is a *run module* if and only if

1. all places and all transitions are labeled,
2. at each place of R at most one edge begins and at most one edge ends,
3. no edge sequence forms a cycle,
4. *R contains all places where no edge ends,
5. R^* contains all places where no edge begins.

The following properties follow directly from this definition; nevertheless, they are important and motivate the concept of run modules.

Theorem 4.2 *1. Each step module is also a run module.*
2. The composition of run modules generates again a run module.
3. Each finite run module R can be composed as $R = P_1 \bullet \ldots \bullet P_n$ from step modules $P_1, \ldots, P_n$.

Proof. 1. By construction, a step module is a run module with one transition.
2. Let P and Q be run modules. The elements of $P \bullet Q$ are elements of P and Q and matches of P^* and $^*Q$. The edges of $P \bullet Q$ are edges of P and of Q that are matches of P^* and $^*Q$. In $P \bullet Q$, two elements of P and Q are connected by at most one path, including one match of P^* and $^*Q$.
3. Let $t_1, \ldots, t_n$ be the transitions of R; let $i < j$ if and only if $t_i < t_j$ in R. Then there exist step modules P_i with transitions t_i $(i = 1, \ldots, n)$ such that $R = P_1 \bullet \ldots \bullet P_n$.
□

Formally, a run module R can be conceived as a partially ordered set: an element x is smaller than an element y, written $x < y$, if there is a sequence of edges from x to y. $^*R$ contains the minimal elements, R^* the maximal elements. The third requirement of Definition 4.4, page 45, excludes elements x, y with both $x < y$ and $y < x$. However, it may very well be the case that neither $x < y$ nor $y < x$. This order defines *causality*; Section 4.7, page 48, details this aspect.

4.5 Expanding the example

The reader may doubt the usefulness of the concept of runs, claiming that e.g. the *bakery* run just describes four event occurrences in a row, as suggested in Equation 4.2, page 45. However, the detailed respect for states pays off when it comes to the investigation of properties of runs. For example, one may extend the bakery run of Figure 4.4, page 45, by the occurrence of a second bake step. Intuitively, the second pastry can be baked immediately after the state *ready to bake* has been reached again, i.e. after the occurrence of *supply to aide*. The first pastry's *move to shop* is not a prerequisite for baking the second pastry. Vice versa, baking the second pastry is not a prerequisite for moving the first pastry to the shop and selling it: baking the second pastry occurs causally independently of moving the first pastry to the shop and selling it.

Figure 4.5 covers this intuition: the second bake module extends the bakery run by a second, bold-faced *bake* transition. Independence of event occurrences is visible as absence of edges between the corresponding transitions. Consequently, a sequence of edges – a *path* in graph theoretic terminology – from an element a to an element b indicates that a occurs causally before b; occurrence of a is necessary for b to occur. Even if, say, the second pastry is baked long before the first pastry is sold, no edges link these event occurrences.

We expand the example even more, by completing the lifeline of the second pastry, as in Figure 4.6. The second *supply to aide* requires not only the second pastry on the counter, but also the aide be free the second time. Hence, *supply to aide* occurs

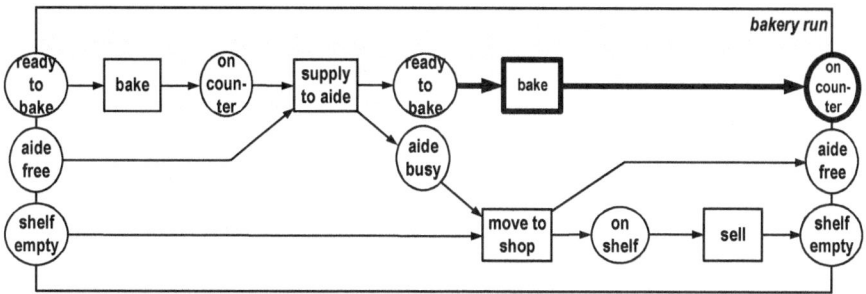

Fig. 4.5 Extending the bakery run by a second *bake* module (bold faced). The composed module *bakery run • bake* extends the bakery run as intuitively expected.

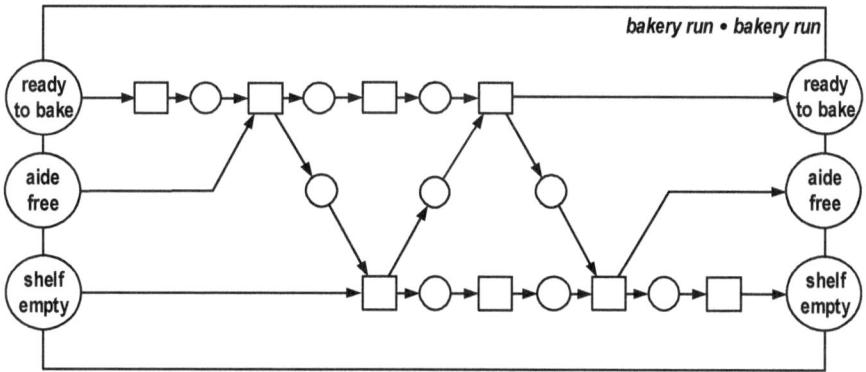

Fig. 4.6 Two composed bakery runs. Two cycles of baking: *bakery run • bakery run*. The elements of the first and the second baking cycle are subtly ordered.

the second time only after the first *move to shop*. Likewise, *move to shop* occurs the second time only after the first pastry has been sold. The sophisticated relationship between the lifelines of the first and the second pastry is formally captured by composing two instances of the bakery run of Figure 4.4, page 45:

$$(2)\text{-}bakery\ runs =_{\text{def}} bakery\ run \bullet bakery\ run \qquad (4.3)$$

To conclude this chapter on runs, it is now obvious how to compose *any* number of baking cycles: For a number n, we denote the composition of n bakery cycles by (n)-*bakery runs*. In particular, for $n = 1$,

$$(1)\text{-}bakery\ runs =_{\text{def}} bakery\ run\,. \qquad (4.4)$$

Now, for $n > 1$, assume the module (n)-*bakery runs* is given. Then we get the module $(n + 1)$-*bakery runs* just by augmenting the module *bakery run*:

$$\text{for } n > 1, \text{ let } (n+1)\text{-}bakery\ runs =_{\text{def}} (n)\text{-}bakery\ runs \bullet bakery\ run\,. \qquad (4.5)$$

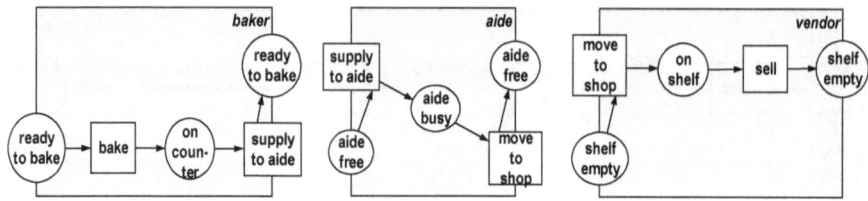

Fig. 4.7 Behavior of the three staff. Most important, their composition *baker • aide • vendor* is identical to the bakery run of Figure 4.4, page 45.

4.6 The bakery run composed from the staff's behavior

Figure 2.5 in Section 2.2, page 14, shows abstract modules of the three bakery staff, the *baker*, the *aide*, and the *vendor*. Figure 4.7 refines the modules, describing each staff member's behavior. Here we see examples of mixed interfaces of net modules: an interface may contain places as well as transitions. Most interesting as well as important for a proper understanding of the bakery model: The composition of the three staff modules yields the *bakery run* of Figure 4.4, page 45:

$$bakery\ run = baker \bullet aide \bullet vendor. \tag{4.6}$$

It is interesting to compare the two quite different approaches to the bakery: The four activities of Figure 2.1 of Section 2.2, page 14, come with interfaces that contain only states. Composition of modules along states is intuitively quite clear. The *three staff* of Figure 2.5, page 16, are composed in Figure 2.6, page 17, along their events.

4.7 A run is ordered by causality

Many models of discrete behavior describe a single run – course of events, progression, system evolution, et cetera – as a sequence of occurrences of states, events, or steps. This is intuitively motivated by the assumption of a global time scale, onto which occurrences of states or events are mapped.

Orders

Definition 3.2, page 20, in Section 3.1 already introduces a notion of order. The literature discusses various variants of orders, called *strict, total, partial, weak, reflexive*, et cetera. Most common are reflexive, total orders, written ≤; most prominent is the order of natural numbers. In contrast, HERAKLIT frequently employs strict, partial orders.

Generally, an order < on a set A is a binary relation on A, i.e. an element $a \in A$ may be smaller than an element $b \in A$, in which case we write "$a < b$".

Definition 4.5 (order) Let A be a set. A binary relation $<$ on A is an *order on A*, if for all $a, b, c \in A$ holds:

- if $a < b$ and $b < c$ then $a < c$ ("transitive"),
- not $a < a$ ("irreflexive").

According to this definition, an order is *asymmetric*: for all $a, b \in A$ holds: if $a < b$ then not $b < a$. An order as defined here is sometimes called *strict*, in order to emphasize *irreflexivity*. An order $<$ is *total* if additionally holds: for all $a, b \in A$: either $a < b$ or $b < a$ or $a = b$. Sometimes we add the adjective "partial" to emphasize that the order in question is *not* total. Every order induces a relation of *unorder* or *independence*.

Definition 4.6 (unorder, independence, #) Let $<$ be an order on A. The *unorder* or *independence* relation # of $<$ is a binary relation on A, defined for all $a, b \in A$ by a # b if and only if neither $a < b$, nor $b < a$.

Notice that according to this definition, unorder is *symmetric*: for all $a, b \in A$ holds: a # b if and only if b # a.
The independence relation of an order $<$ can be *transitive*. In this case, $<$ is a *weak order*.

Definition 4.7 (weak order) An order $<$ on a set A is a *weak order*, if for all $a, b, c \in A$ holds: If a # b and b # c then a # c.

We frequently employ general, *non-weak*, *partial* orders.

In contrast, HERAKLIT orders event occurrences not by a *temporal* before-after relationship, but by a *causal* one. This concept is favorable in many aspects; it will come with local instead of global states, and causally independent event occurrences. This setting is particularly useful to cover subtle aspects of the composition of runs. With the above Definition 4.4, page 45, of run modules and the above Theorem 4.2, page 46, in a run, an element a can be considered as a causal prerequisite for an element b, if there exists a sequence of edges from a to b: evolution is triggered by *causality*. In this setting, causality is a *partial* order.

Definition 4.8 (causal order) Let $R = (P, T; E)$ be a run module, let $a, b \in P \cup T$. Then *a is causally ordered before b* in R, written $a < b$, if and only if there exists a sequence of edges from a to b.

For example, the module *bakery run* in Figure 4.4, page 45, contains two occurrences of the state labeled *ready to bake*. Four edges connect them via *bake*, *on counter*, and *supply to aide*. So, the instance of *ready to bake* on the left interface is causally ordered before the instance on the right interface. The state *aide free* is reached causally before the event *supply to aide* occurs, as well as before the state *aide busy* is reached, and before *ready to bake* is reached the second time.

Theorem 4.3 Causal order *is in fact an* order.

Proof. To prove this theorem, observe that a path v from a to b, and a path w from b to c, generates a path vw from a to c. Furthermore, no edge sequence of a run module forms a cycle; hence not $a < a$. □

We consider this order in more detail, studying in particular sets of mutually or pairwise unordered elements. Causal order is a partial order in the previous box 4.5, page 49.

Definition 4.9 (independence, #) Let $R = (P, T; E)$ be a run module, let $<$ be its causal order. Two elements $a, b \in P \cup T$ are *independent*, or *concurrent*, written $a \# b$, if neither $a < b$, nor $b < a$.

Causal and temporal order in the digital world

The literature describes a plethora of conceptualizations, definitions, and methods for studying causality [16, 28, 49, 67, 87, 96]. It is even often argued that understanding a composed system implies identifying the causal relationship among its components. A closer look at the above contributions reveals an important premise: causality is implicitly or even explicitly based on a *discrete* or *continuous* time axis, represented by *natural* or *real* numbers. HERAKLIT refrains from such a premise and understands causality as a partial order of events in the digital world [26, 68, p. 36f.].

Three aspects relate temporal and causal order:

1. Every causal relation induces a temporal order: The cause precedes its effect. That event a happens in time before event b is a necessary condition for a to cause b. On the other hand, that a happens in time before b is not a sufficient condition for the causal order of the two events.
2. Every causal relation can be mapped onto different temporal orders. In other words, different observers of two events may observe the two events in different temporal orders.
3. A temporal order is intuitively *weakly ordered*: If two events are not ordered in time, they appear intuitively at the *same time*; i.e. they occur *simultaneously*. Intuitively, the property of simultaneity is transitive. In sharp contrast, the relation of causal independence of events is in general not transitive.

[27] provides a deeper discussion of different temporal models and their causal understanding.

Generally, by construction, the gates in the left interface $^*R$ of a run R are pairwise independent, as are the gates of R^*. For a more general example, in Figure 4.8, each vertex of the small shaded ellipse B is concurrent with each vertex of the large shaded ellipse A.

It is important to observe that independence is not transitive, i.e. a run may contain elements a, b, c with $a \# b$ and $b \# c$, but not $a \# c$. Figure 4.8 shows an example.

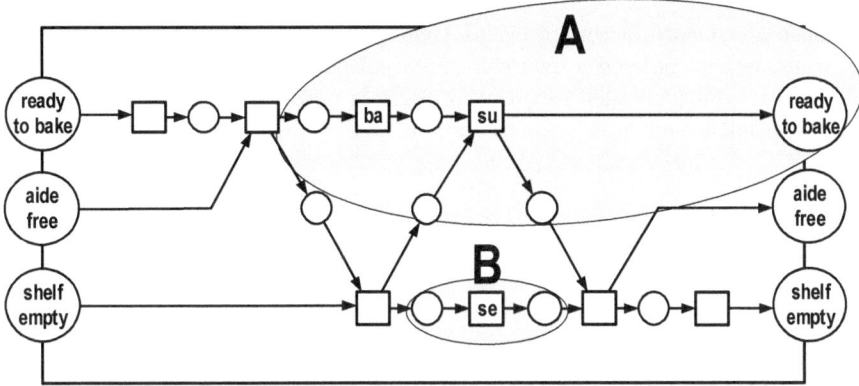

Fig. 4.8 Each vertex in A is concurrent with each vertex in B. For example, within A and B, ba # se, and se # su, but $ba < su$. Hence, # is not transitive. As every temporal order is transitive, # cannot be conceived as a temporal order.

Theorem 4.4 *The relation # on a run is in general* not *transitive.*

Proof. To prove this theorem, Figure 4.8 shows an example: ba # se and se # su, but $ba < su$.   □

This observation implies that # cannot intuitively be conceived as "at the same time" or "simultaneously": equality *in time* is always transitive. This contradicts the above definition of the relation "#", which is intransitive, in general.

> **Interleaved runs**
>
> Nearly all approaches to model discrete behavior describe a single run as a sequence of events or steps, with the events to occur totally ordered "along the passing of time". Time is usually assumed to be "given", mostly without referring to a specific clock. Sometimes, more than one event may be modeled to occur at the same time: At a point in time a set of events occurs. A run is then a sequence of sets of events. Events or sets of events are said to be *interleaved* in this setting. As an example, the run of the bakery as in Figure 4.5, page 47, has five representations as interleaved runs:
>
> 1. {bake}, {supply to aide}, {bake}, {move to shop}, {sell};
> 2. {bake}, {supply to aide}, {bake, move to shop}, {sell};
> 3. {bake}, {supply to aide}, {move to shop}, {bake}, {sell};
> 4. {bake}, {supply to aide}, {move to shop}, {bake, sell};
> 5. {bake}, {supply to aide}, {move to shop}, {sell}, {bake}.
>
> The first, third, and fifth representation can be conceived as sequences of events, i.e. totally ordered sets of events. The second representation can be conceived as a partial order, with *bake* and *move to shop* unordered. Likewise, in the fourth representation, *bake* and *sell* are unordered. Generally,

such a sequence of sets of events yields a *weak order*. In the context of runs, weakly ordered events can be conceived to occur in *lockstep*. The complement ("unorder") of a weak order is transitive; in contrast to the independence relation #, as introduced in Definition 4.9, page 50.

Chapter 5
States and events: elementary system modules

We now proceed from the level of single runs to the level of *systems*: a system describes a set of runs. The central idea: a system does not describe single occurrences of states and events – as runs do – but the states and events themselves.

5.1 The notion of elementary system modules

Technically, an *elementary system module* is a net module, see Section 4.2, page 43. Its places are *states*, and its transitions are *events*. In contrast to a run, an elementary system module may include edge sequences that form alternatives or cycles.

To represent a set R of runs, an elementary system module M represents all occurrences of a state p in any of the runs of R by the state p itself, as a place of M. Likewise, M represents all occurrences of an event e in any of the runs of R by the event e itself, as a transition of M.

Definition 5.1 (elementary system module) An *elementary* HERAKLIT *system module* is a HERAKLIT net module $M = (P, T; E)$ where each place $p \in P$ represents a *state* and each transition $t \in T$ represents an *event*.

Graphical convention: Some states may be distinguished as *initially taken*; they are inscribed with a black dot.

Figure 5.1 shows three elementary HERAKLIT system modules. Figure 5.2 composes them.

Each elementary HERAKLIT system module characterizes an – in general infinite – set of runs, as defined in the following section.

P. Fettke, W. Reisig, *Understanding the Digital World*,
https://doi.org/10.1007/978-3-031-61898-7_5

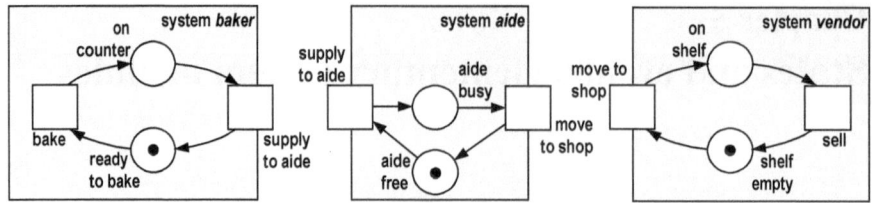

Fig. 5.1 Three HERAKLIT system modules. Each of the three modules describes a cyclic behavior, with two alternating transitions: the *baker system* with *bake* and *supply to aide* (left); the *system aide* with *supply to aide* and *move to shop* (middle); the *system vendor* with *move to shop* and *sell* (right).

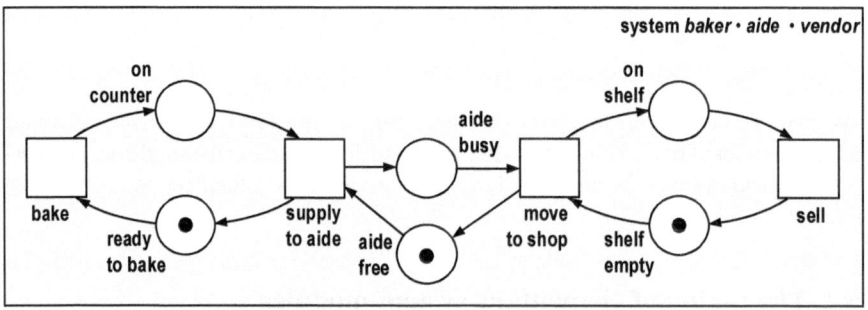

Fig. 5.2 Composing the three HERAKLIT modules of Figure 5.1. The three cycles in that Figure synchronize along the two transitions *supply to aide* and *move to shop*.

5.2 Steps and runs of elementary system modules

In Section 4.3, page 43, a step module M has been defined as a labeled net module with one transition. If the labeling describes a $1 - 1$-mapping onto an elementary net module N, we conceive M as a defining *step of N*.

Definition 5.2 (step of an elementary system module) Let $M = (P, T; E)$ be an elementary system module, and let $N = (Q, \{u\}; B)$ be a step module. Let l be a labeling of $Q \cup \{u\}$. Then N is a *step of M* if and only if:

1. u is labeled by some $t \in T$;
2. (b, u) is an edge of N if and only if $(l(b), t)$ is an edge of M;
3. (u, b) is an edge of N if and only if $(t, l(b))$ is an edge of M.

Examples include the four step modules of Figure 4.2, page 44: They are steps of the elementary step module of Figure 5.2. Based on the definition of run modules in Section 4.4, page 44, a run module of an elementary net module N is composed from steps of N:

Definition 5.3 (run of elementary system module) Let $M = (P, T; E)$ be an elementary system module, and let R be a run module. R is a *run of M* if and only if each step of R is a step of M.

For example, the run in Figure 4.4, page 45, is a run of the module in Figure 5.2, page 54.

Later on, places of an elementary system module will be understood, in a mathematical framework, as propositions with varying truth values.

The definition of run modules R on net modules M follows the locality principle of HERAKLIT, i.e. the absence of any global assumption. In particular, a system module is not equipped with an initial state. This gives rise to important properties: Firstly, any two runs can be composed, returning again a run. Secondly, a run of a system is also a run of each "bigger" system.

Theorem 5.1 *The composition of runs of an elementary system module M is again a run of M.*

Proof. To prove this theorem, notice that in Section 4.5, page 46, we proved already that the composition of $P \bullet Q$ of run modules of P and Q generates again a run module. Let P and Q be steps on M. The steps of $P \bullet Q$ are steps of P and steps of Q. Hence, the steps of $P \bullet Q$ are steps of N. □

Theorem 5.2 *Let L, M, and N be elementary net modules. Let R be a run module of M. Then R is a run module of $L \bullet M \bullet N$.*

Proof. To prove this theorem, let P be a run of M. The steps of P are steps of M. The steps of M are steps of $L \bullet M \bullet N$. □

Nevertheless, occasionally, we focus on runs that start with a distinguished set Q of states. In such cases, one considers only runs R that are *commenced*, i.e. Q is the set of labels of $^*R$. In graphical representations, a dot decorates each element of Q. As an example, with respect to Equation 4.5, page 47, for each $n \in \mathbb{N}$ holds:

$$(n)\text{-bakery runs is a commenced run of the module in Figure 5.2, page 54.} \quad (5.1)$$

Interleaving semantics

An elementary system module defines a set of *interleaved runs* in an obvious manner. For example, the five interleaved runs in the box *interleaved runs* at the end of Chapter 4, page 51, are runs of the system module of Figure 5.2, page 54. In fact, interleaved runs have dominated and still dominate the study of runs of Petri nets. They are easy to manage, just words over the alphabet that consists of the transitions. The interleaved runs of an elementary system module M generate a transition system. This in turn can be analyzed by means of temporal logic, thus proving important properties of M. However, this framework, frequently called as *interleaving semantics*, comes with some disadvantages: interleaving semantics assumes global states, and each step connects two global states. This is intuitively not convincing. And composition of modules implies technically awkward definitions of corresponding composed interleaved runs of the composed system.

Within the HERAKLIT framework, handling of partially ordered runs is no
longer technically complicated.

5.3 Conflicting situations

Here we consider a variant of the bakery system, now including two shops. To this
end, the modules of Figures 5.1, page 54, and Figure 5.2, page 54 are extended, as
in Figure 5.3: the *system vendor* is complemented by *system vendor'*, with primed
names of places and transitions. The aide system's right interface is extended by the
transition *move to shop'*. Technically, two arcs start at the *aide busy* place of *system
aide'*. In a situation where *aide busy* has been reached, the two steps *move to shop*
and *move to shop'* both may occur. But they conflict with each other: After one of
them occurs, the other one can no longer occur.

Figure 5.4 shows the composition of the system modules of Figure 5.3:

$$two\ shop\ system =_{\text{def}} bakery \bullet aide' \bullet vendor \bullet vendor' \qquad (5.2)$$

If the three states *aide busy*, *shelf empty*, and *shelf empty'* are reached, the two
steps *move to shop* and *move to shop'* both may occur. But only one of them will
occur. This gives rise to many different runs. For example, the run in Figure 4.6,
page 47, is a run of the *two shops system*: the upper vendor of *two shops systems* is
served twice. But the run in Figure 5.5 is also a run of this system: serving the upper

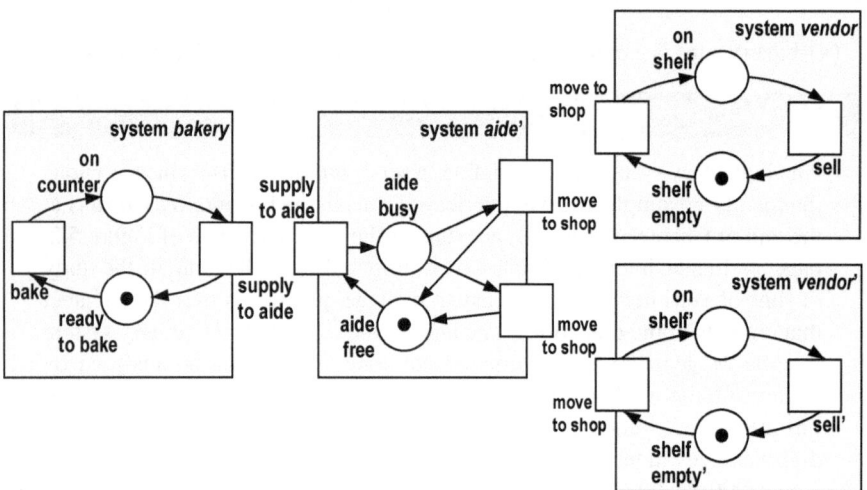

Fig. 5.3 The *system aide'* is non-deterministic: in a situation where *aide busy* has been reached,
the two steps *move to shop* and *move to shop'* both may occur. But they conflict with each other:
After one of them occurs, the other one can only occur after the next occurrence of *supply to aide*.

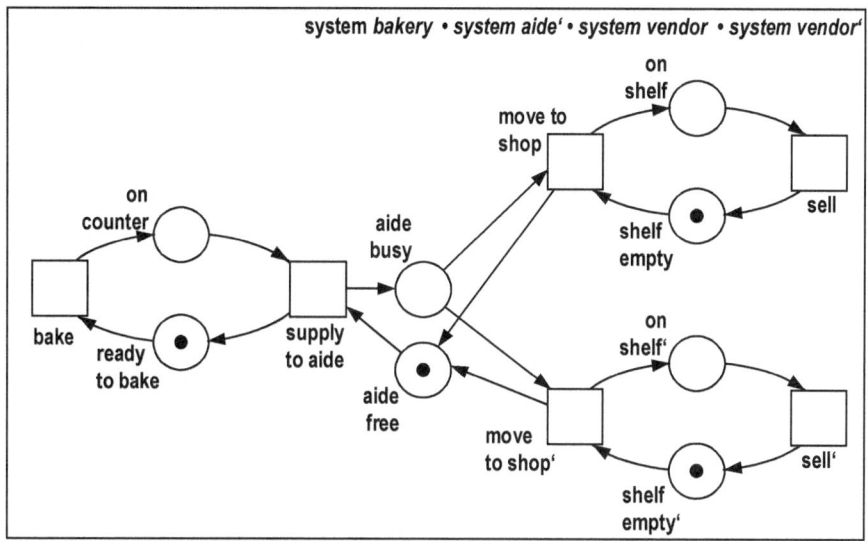

Fig. 5.4 Composition of the two shops modules. The alternative between the transitions *move to shop* and *move to shop'* gives rise to *infinitely* many run.

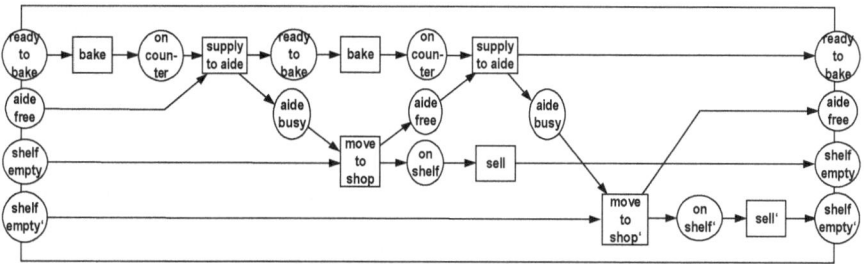

Fig. 5.5 The two vendors are served. *System vendor* is followed by *system vendor'*.

vendor is followed by serving the lower vendor. Nevertheless, the lower vendor may finish, in a temporal setting, before the upper vendor. Altogether, this elementary system module describes infinitely many runs.

Such conflicting situations are not undesirable per se: a system model may purposely leave an alternative undecided. Nevertheless, one may wish to resolve such conflicting situations. To this end, HERAKLIT allows us to formulate rules to resolve conflicts. As an example, the module *alternating aide system* of Figure 5.6 generates alternating service for the two shops, starting with the upper one.

Confusion

In the elementary system module of Figure 5.4 a run that has reached the state *aide busy* may continue alternatively with one of the two transitions *move to shop* and *move to shop'*: The state *aide busy* is the starting point of the two transitions, so they are in *conflict*. In general, conflict may depend not on just one, but on many states, even many independently reached states. Then the notion of conflict is confused; there is no objective criterion to decide whether or not a conflict has occurred. Figure 5.7, page 59, shows an example.

5.4 Equally labeled gates

Here we want to specify more general composition facilities in HERAKLIT models. More concretely, it should be an easy task to swap the two vendor modules, *system vendor* and *system vendor'* in the above example, with *system vendor'* to be served first; instead of *system vendor*, as in Figure 5.4, page 57. To this end, we label the two transitions in the *alternating aide system'* identically, as in Figure 5.8. Additionally, also the transition in the left interface of *system vendor''* is labeled alike.

As stated in Section 3.1, page 20, in this case we assume an order, top-down, on the *move to shop* labeled transitions of the right interface of the module *alternating aide* of Figure 5.6. According to the definition of composition of modules in Section 3.3, page 24, in the module

$$bakery \bullet alternating\ aide' \bullet vendor \bullet vendor'', \qquad (5.3)$$

the upper *move to shop* labeled transition of *alternating aide system* merges with the *move to shop* labeled transition of *system vendor*, and the lower *move to shop* labeled transition of *alternating aide system* merges with the *move to shop* labeled transition of *system vendor''*.

Swapping the two vendor systems yields the module

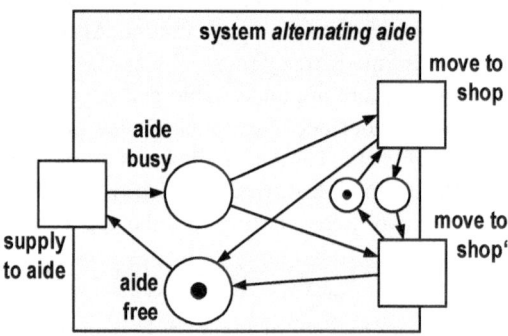

Fig. 5.6 Alternating service for the two shops. The transitions labeled *move to shop* and *move to shop'* are linked by a *synchronizing circuit* that alternately enables the two transitions.

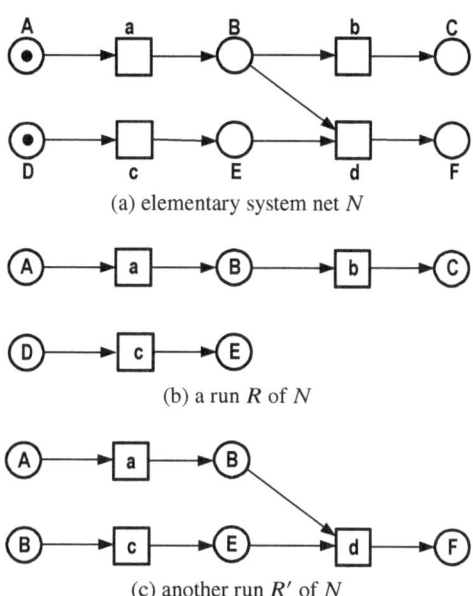

(a) elementary system net N

(b) a run R of N

(c) another run R' of N

Fig. 5.7 If in the run R, the transitions a and b occur "quickly" and c occurs "slowly", the state B is abandoned before state E has been reached. In this case, the transition d has no chance to occur. The transitions b and d do not conflict. If the transition a occurs "slowly" and c occurs "quickly", the states B and E are eventually reached "at the same time". In this case, b and d are conflicting. A run comes without any notion of time and speed of transition occurrences. So, "conflict" remains a *confused* notion in the system net N.

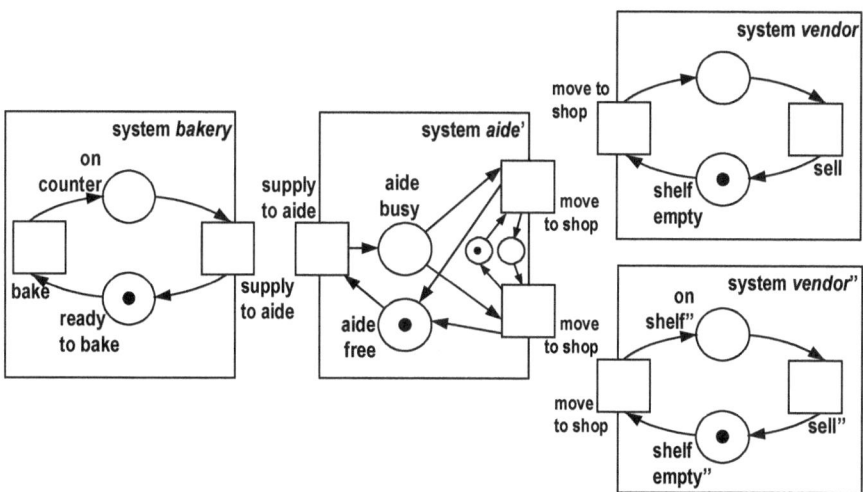

Fig. 5.8 Interfaces with identically labeled gates. The label *move to shop* is the label of both transitions in the right interface of the *system aide*, as well as in the left interface of *both* vendor systems. The assumption of a top-down order merges the *system vendor* with the upper transition, and the *system vendor"* with the lower transition.

$$bakery \bullet alternating\ aide' \bullet vendor'' \bullet vendor\ . \qquad\qquad (5.4)$$

Here, the upper *move to shop* labeled transition of *alternating aide system* merges with the *move to shop* labeled transition of *system vendor''*, and the lower *move to shop* labeled transition of *alternating aide system* merges with the *move to shop* labeled transition of *system vendor*.

This example shows that HERAKLIT can represent subtle behavioral aspects most adequately. Furthermore, the modules of the Equations (5.3) and (5.4) show that the abstract representation of modules as a composition of given modules, just by use of the symbol "•" for the composition operator, complements the diagrammed representation in a particularly useful way.

Related work to Part II

Many modeling frameworks base their behavioral and dynamic aspects on *transition systems*. Essentially, a transition system is a directed graph. Each node represents a state; frequently, one state is distinguished as "initial". Each edge $p \xrightarrow{t} q$ of a transition system represents a *step from state p to state q*, triggered by the *event t*. A *run* is then a path in the transition system, starting at the initial state.

The states of a composed transition system $A \bullet B$ are usually the tuples (a, b) of states a of A and b of B. This leads to an "explosion" of the number of states, steps, and runs. For instance, composition of n transition systems, each with k states, yields a composed transition system with k^n states and k^{2n} steps, and k^{2nm} runs of length m.

Modeling frameworks such as statecharts [35], process algebras [57, 58], et cetera assume abstract events, abstracting away from concrete data, in order to keep the formalism better manageable. But they are based on transition systems, and as outlined above, the number of states, steps, and runs of those transition systems quickly "explodes". To capture this problem, the calculi provide means to represent "big" transition systems. For example, statecharts extend state machines by hierarchical states, parallel states, and history dependent states. Process algebras suggest algebraic calculi, with parallel composition that semantically generates a transition system as discussed above. This composition operator is associative, at the price of non-determimism.

Dynamic behavior is often modeled by Petri nets [71, 76]. Some versions of Petri nets are essentially just transition systems. The concept of *runs*, as in Chapter 4, page 41, was suggested by *Carl Adam Petri* in [70], and they became known as "partial order" or "distributed runs", in contrast to "interleaved runs" of transition systems. [46] suggests a compositional partial order semantics that might also be applied in HERAKLIT. But Kindler's composition operator just fuses equally labelled interface elements – and is not associative. [11] compiles properties of infinite partial order runs.

Partial order runs occur also outside Petri nets. The simplest form is a *message sequence chart* (MSC). Such a chart is a set of sequential runs, where a run may send or receive messages to or from other runs. Among the many versions of this idea, a particularly refined one is described in [36].

For decades, special classes of Petri nets have been suggested, to capture important classes of applications, to gain better analysis techniques, and to achieve useful notions of composition. In one of many typical examples, [9] considers nets that model workflows, and composes them in a way that preserves the important property of *soundness*. For nets with data, [18] composes a special kind of HERAKLIT-like modules, allowing the composition of place invariants. [10] combines ideas of Petri nets and process algebras.

Though firmly based on Petri nets, HERAKLIT re-considers some technical and conceptual standards. First of all, fundamental for HERAKLIT is the notion of a module. The usually defined "firing rule" of a transition t is replaced by event occurrences of t. A partially ordered run is not necessarily assigned to a system net, and HERAKLIT refrains from initial states.

HERAKLIT applies the composition calculus of Part I to Petri nets. In particular, the composition calculus yields simple and convenient technical tools for partially ordered runs.

From an application perspective, many different approaches to model dynamics have emerged over the years. Event-driven process chains blossomed since *SAP* started to document their products for enterprise resource planning using this approach [83, 91]. Various contributions try to overcome the lack of formal foundations of event-driven process chains [47, 54, 92]. Typically, Petri nets are used as a formal basis. Similarly, today's de-facto standards for modeling dynamics, such as UML activity diagrams and BPMN, borrow ideas from Petri nets to describe dynamics.

Part III
The statics pillar

After the architecture pillar in Part I and the dynamics pillar in Part II, we now turn to the third HERAKLIT pillar, covering real- or imagined-world items and data, as they occur in computer-integrated systems. Chapter 6 suggests a systematic way to cope with such items and data. Chapter 7 deals with their symbolic representation in HERAKLIT system models.

Caveat. Contentwise, this part presents standard material of predicate logic and algebraic specifications, as they are used in informatics. At first reading, the examples of the figures as in the subsequent Part IV may provide sufficient intuitive understanding. In particular, notions such as functions, predicates, relations, *or* propositions, *are sufficiently clear to many readers who are already familiar with predicate logic or algebraic specifications. This likewise applies to terms that include variables, and contribute to symbolic representations of real-world items. We nevertheless present a systematic built-up of those notions, in order to remain self-contained, and to demonstrate the close relationship between HERAKLIT and predicate logic.*

Chapter 6
Real- or imagined-world items and data

HERAKLIT describes items and data in a systematic way, based on classical notions of sets. The next Section 6.1 systematically derives the relevant concepts from very elementary notions. After that, Section 6.2 presents the well-established framework of *structures* for those concepts.

6.1 Domains

A computer-integrated system is in general concerned with lots of real- or imagined-world items and data. Some of those items or data are specific to the system under consideration. We collect those items and data in *domains*. For example, the bakery system includes the items *bread*, *cake*, and *pie*. These items constitute the domain of pastries of the system. Later on, the organization of behavior of the bakery system requires a *description* of the pastries to be baked. This is achieved by the data *"bread"*, *"pie"* and *"cake"*. These data are the elements of the domain of *descriptions* of the system. Notice the difference between a pastry and its description. This is a fundamental difference: formulated vividly, *cake* you can eat, but *"cake"* you can only read.

In the context of HERAKLIT system models, a set of real- or imagined-world items and data is called a *domain*. New domains can be derived from given domains as subsets, tuple sets, relations, and functions.

Definition 6.1 (derived domains)

- A *domain* is a set of *items*.
- Each *subset B* of a domain *A* is a domain; i.e. each element of *B* is also an element of *A*.
- For domains $A_1, \ldots, A_n$, the *Cartesian product* $A_1 \times \cdots \times A_n$ is a domain; each element of $A_1 \times \cdots \times A_n$ is a tuple $(a_1, \ldots, a_n)$, with $a_1 \in A_1, \ldots, a_n \in A_n$.
- For a domain *A*, each *multi-set* over *A* is a domain; a multi-set contains each element any number of times.

© The Author(s), under exclusive license to Springer Nature Switzerland AG 2024
P. Fettke, W. Reisig, *Understanding the Digital World*,
https://doi.org/10.1007/978-3-031-61898-7_6

- For domains $A_1, \ldots, A_n$, a subset of $A_1 \times \cdots \times A_n$ is frequently called a *relation over* $A_1 \times \cdots \times A_n$.

For example, in a variant of the bakery example, the baker may bake pastries not one by one, but on trays, with each tray containing multiple instances of *bread*, *pie*, and *cake*. The pastries on a tray constitute a multi-set.

In a systematic build-up, domains are mutually related by *functions*. For example, $f : pastries \to \mathbb{N}$ is a function that assigns each pastry p its price $f(p)$. As another example, the function *des* assigns each pastry its description.

To remain consistent, complete, and self-contained, we conceive functions as special relations; predicates are special functions, and propositions are special predicates.

Definition 6.2 (functions, predicates, propositions) Let $M_1, \ldots, M_{n+1}$ be domains.

1. A *function* f is a relation over $M_1 \times \cdots \times M_{n+1}$ such that for each tuple $(m_1, \ldots, m_n) \in M_1 \times \cdots \times M_n$ there exists exactly one element m_{n+1} such that $(m_1, \ldots, m_{n+1}) \in f$. Then f is written:

$$f : M_1 \times \cdots \times M_n \to M_{n+1}, \tag{6.1}$$

and m_{n+1} is written:

$$f(m_1, \ldots, m_n). \tag{6.2}$$

2. A *predicate* p of type $M_1 \times \cdots \times M_n$ is a function of the form $p : M_1 \times \cdots \times M_n \to \{true, false\}$. Such a predicate p is said to *hold for*, or to *apply to*, a tuple $(m_1, \ldots, m_n) \in M_1 \times \cdots \times M_n$, if $p(m_1, \ldots, m_n) = true$. $(m_1, \ldots, m_n)$ is an *instantiation* of p. We write $p : M_1 \times \cdots \times M_n$ to indicate the *type* of p.

3. A *proposition* q is a predicate shaped $q : \{a\} \to \{true, false\}$. A proposition q is said to *hold*, if $q(a) = true$. Choice of the item a is entirely irrelevant, and is not mentioned in the context of propositions.

> ### World and language – manual work and mouth work
>
> A loaf of bread can be eaten, but its description cannot be eaten. A customer can sit in the waiting room, but cannot be stored in a database. Instead, information about a customer, his *description*, can be processed by a database system. And a vending machine requires coins to be inserted before an item can be purchased. Before an item can be purchased, the machine typically provides information about the item, such as its brand label, weight, and price. After the required coins are inserted, the machine should dispense the desired item that matches the announced information about it.
>
> All of these examples show that there is a distinction between the *world* and its *description*. This difference can be vividly recounted by comparing work done with the *hand* to work done with the *mouth*. Traditionally, a computer only performs mouth work, such as storing customer data, sorting informa-

tion, calculating revenues. But in a digital world, manual work and mouth work are closely intertwined; cyber-physical systems emerge, e.g. software controls an autonomous vehicle, a production machine, or a logistical process. Nevertheless, the recounted difference is of general importance for understanding the world, as the work of Peter Janich [42] especially focuses on. In particular, HERAKLIT provides concepts for dealing with items *and* with data in the digital world.

6.2 The world in good shape: structures

HERAKLIT organizes items and data on the basis of domains. We distinguish four kinds of domains of a HERAKLIT module:

- *specific* domains, such as the *pastries* and the *descriptions* as in our running example,
- two *standard domains*:

$$\text{the natural numbers } \mathbb{N}, \text{ and the } truth\ values\ \{\text{true, false}\} \qquad (6.3)$$

- *derived* domains, as defined in Section 6.1, page 67,
- *inherited* domains, provided by other modules in the *wrapper* of the module under consideration.

In addition to domains, also functions are a fundamental concept of static aspects of modules. For example, the function *des* : *pastries* → *descriptions* assigns each pastry its description. Vice versa, the function *item* : *descriptions* → *pastries* assigns each description of a pastry the pastry itself. Both functions bridge the gap between real-world items, e.g. pastries, and a digital representation.

The function *next* : *descriptions* → *descriptions* organizes a cyclic structure on the pastries: it will be used to inform the baker which pastry to baked next. If *"pie"* is the description of what has been baked last time, the next pastry to be baked is a *bread*, hence, let *next("pie")* = *"bread"*. Likewise, let *next("bread")* = *"cake"*, and *next("cake")* = *"pie"*.

In addition to domains and functions, finitely many items from some of the domains are identified as *constants*. Usually, many items of domains can be addressed by repetitive application of functions to constants.

For example, of the six elements of the two domains *pastries* and *descriptions*, we take only *"pie"* as a constant. The other five elements can (and will) be addressed via the structure's functions. As a typical case,

$$bread = item(next(des(pie))). \qquad (6.4)$$

As a general notion, for a domain A,

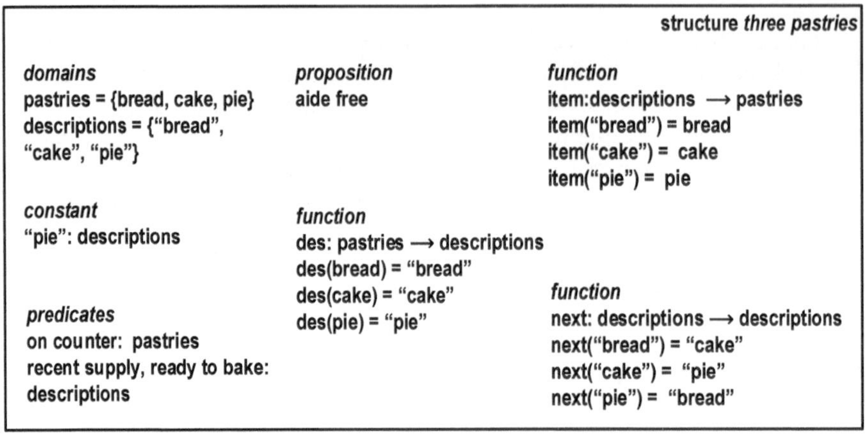

<table>
<tr><td colspan="3" align="right">structure three pastries</td></tr>
</table>

domains	*proposition*	*function*
pastries = {bread, cake, pie}	aide free	item:descriptions ⟶ pastries
descriptions = {"bread",		item("bread") = bread
"cake", "pie"}		item("cake") = cake
		item("pie") = pie
constant	*function*	
"pie": descriptions	des: pastries ⟶ descriptions	
	des(bread) = "bread"	
	des(cake) = "cake"	*function*
predicates	des(pie) = "pie"	next: descriptions ⟶ descriptions
on counter: pastries		next("bread") = "cake"
recent supply, ready to bake:		next("cake") = "pie"
descriptions		next("pie") = "bread"

Fig. 6.1 A structure for three pastries.

$$\text{an element } c \in A \text{ is a } \textit{constant of type A.} \tag{6.5}$$

This completes the static, not-changing aspects of HERAKLIT modules.

Changing, dynamic aspects of modules have been presented in Part II, concentrating on aspects that do not depend on data. There, a state is reached or abandoned by the occurrence of an event. Formally, such a state s is a proposition that starts holding when s is reached, and stops holding when s is abandoned.

In real or imagined systems, behavior very well depends on items and data. In this case, behavior will accordingly be described by means of predicates.

This brings us to the notion of *structures*. In predicate logic, such structures are denoted – and serve as – *models*; sometimes they are also called *Tarski structures*. They must not be confused with architectures, as discussed in Part I, page 11.

Definition 6.3 (structure) A *structure S* consists of:

- a finite set D of domains;
- finitely many functions, each of the form $f : M_1 \times \cdots \times M_{n+1} \to M_{n+1}$, where each M_i is a domain or a derived domain; the $n + 1$-tuple $(M_1, \ldots, M_{n+1})$ is the *arity* of f;
- finitely many constants and predicates, with domains or derived domains as types;
- finitely many propositions.

As an example, Figure 6.1 shows the structure *three pastries* in the context of our running example. This structure includes two domains, three functions, three predicates, one constant, and one proposition.

As a second example, Figure 6.2 shows a structure with more elements. The two structures are different; nevertheless, they are *similar*.

Definition 6.4 (similar structures) Let S and T be two structures. Then S and T are *similar* if and only if there exists a bijective function Φ between the domains, functions, constants, predicates, and propositions of S and those of T such that

		structure *five pastries*
domains pastries = {bread, cake, pie, roll, biscuit} descriptions = {"bread", "cake", "pie", "roll", "biscuit"}	*proposition* aide free	*function* item:descriptions ⟶ pastries item("bread") = bread item("cake") = cake item("pie") = pie
constant "pie": descriptions	*function* des: pastries ⟶ descriptions des(bread) = "bread" des(cake) = "cake" des(pie) = "pie" des(roll) = "roll" des(biscuit) = "biscuit"	item("roll") = roll item("biscuit") = biscuit
predicates on counter: pastries; recent supply, ready to bake: descriptions.		*function* next: descriptions ⟶ descriptions next("bread") = "cake" next("cake") = "pie" next("pie") = "roll" next("roll") = "biscuit" next("biscuit") = "bread"

Fig. 6.2 A structure for five pastries.

1. for each function $f : M_1 \times \cdots \times M_n \to M_{n+1}$ of S, $\Phi(f)$ is a function of T, of the form $f : \Phi(M_1) \times \cdots \times \Phi(M_n) \to \Phi(M_{n+1})$,
2. for each constant or proposition c of S with type M, $\Phi(c)$ is a constant of T with type $\Phi(M)$,
3. For each predicate p of S with type M, $\Phi(p)$ is a predicate of T with type $\Phi(M)$.

Summing up, a structure assembles real-world items, data, constants, functions, predicates, and propositions that a dynamic system updates, generates, extinguishes, computes, and transforms.

Chapter 7
Do it symbolically:
signatures, terms, and interpretations

A HERAKLIT model includes real- and imagined-world items, such as *pastries*, *descriptions*, *shelves*, et cetera, collected in one or more structures, as defined in the previous Chapter 6. To share a HERAKLIT model among various stakeholders, to systematically update it, et cetera, it is necessary to represent the structure *symbolically*. For the structure *three pastries* of Figure 6.1, page 70, a symbolic representation is easy: Each domain is finite, hence finitely many symbols suffice to represent each domain and each constant, predicate, and function. Each predicate can explicitly be listed, and for each function f, the result $f(a)$ can explicitly be declared for each argument a. However, this fails for a structure with at least one infinite domain. For frequently used domains such as the natural numbers, there are standard conventions for the symbolic representation of elements, e.g. sequences of digits. But domains are often not entirely known. Therefore, elements of such domains cannot be represented symbolically.

To solve this problem, one assigns each structure S a *signature*, Σ, and regains S by the corresponding interpretation of the signature's symbols.

7.1 Symbols for structures: signature

A structure usually comes with a *signature*, providing a symbol for each of the structure's components. Each domain gets its symbol, called its *type*, in analogy to the types of components of structures.

Definition 7.1 (signature) A *signature* Σ consists of:

- finitely many symbols for domains;
- finitely many function symbols, each such symbol generated from domain symbols has a tuple of domain symbols as its *arity*;
- finitely many constant symbols and predicate symbols, each such symbol has a tuple of domain symbols generated from domain symbols as its *type*;
- finitely many symbols for propositions.

P. Fettke, W. Reisig, *Understanding the Digital World*,
https://doi.org/10.1007/978-3-031-61898-7_7

		signature *bakery*
domain symbols pastries descriptions	*predicate symbols* on counter: pastries recent supply, ready to bake: descriptions	*function symbols* des: pastries → descriptions item: descriptions → pastries
constant symbol p: descriptions	*proposition symbol* aide free	next: descriptions → descriptions

Fig. 7.1 A signature for the bakery. This signature provides the symbols to cope with the structures *three pastries* and *five pastries* in Figures 6.1 and 6.2, pages 70 and 71, and many other structures.

In practical contexts, the denotation of a real-world item or a function of a structure S is closely related to a signature's symbols. To avoid confusion, symbols and symbol sequences that refer to a signature will be underlined. In particular, in the context of a signature Σ, we write:

- $\underline{f} : \underline{A_1}, \ldots, \underline{A_n} \to \underline{A_{n+1}}$ for a function symbol $\underline{f}$ of Σ with arity $(\underline{A_1}, \ldots, \underline{A_{n+1}})$;
- $\underline{c} : \underline{A}$ for a constant symbol $\underline{c}$ of Σ with type $\underline{A}$;
- $\underline{p} : \underline{A}$ for a predicate symbol $\underline{c}$ of Σ with type $\underline{A}$.

Figure 7.1 shows a typical signature.

7.2 When signatures and structures fit: Σ-structures

For a given signature Σ, a structure S is a Σ-*structure* if the domains, functions, constants, predicates, and propositions of S correlate with corresponding symbols of Σ.

Definition 7.2 (Σ-Structure) Let Σ be a signature, and let $\underline{A}, \underline{A_1}, \ldots, \underline{A_{n+1}}$ be domain symbols of Σ. Let S be a structure that consists of:

- a domain A_S for each domain symbol $\underline{A}$ of Σ;
- a function $f_S : A_{1_S} \times \cdots \times A_{n_S} \to A_{n+1_S}$ for each function symbol $\underline{f}$ with arity $(\underline{A_1}, \ldots, \underline{A_{n+1}})$ of Σ;
- a constant c_S of type A_S for each constant symbol $\underline{c}$ with type $\underline{A}$ of Σ;
- a predicate p_S with domain A_S for each predicate symbol $\underline{p}$ with type $\underline{A}$ of Σ;
- a proposition p_S for each proposition symbol $\underline{p}$.

Then S is a Σ-*structure*, also called an *interpretation* of Σ.

For example, the structure *three pastries* of Figure 6.1, page 70, is a structure of the signature *bakery* in Figure 7.1, page 74. The structure *five pastries* of Figure 6.2, page 71, is also a structure of the signature *bakery*.

In fact, all interpretations of a signature are similar.

Theorem 7.1 *Two relational structures are interpretations of the same signature if and only if they are similar.*

Proof. To prove this theorem, let S and T be Σ-structures. For a domain, function, constant, predicate, or proposition x of S, let $\Phi(x)$ be a domain, function, constant, predicate, or proposition of T, where x and $\Phi(x)$ are interpretations of the same symbol of Σ. Then S and T are similar. □

7.3 Composing symbols: terms

A signature Σ gives rise to – in general infinitely many – *terms*, viz. composed symbols, intended to represent elements of Σ-structures. To define terms, we start with *variables*. Each variable is assigned a domain symbol.

Definition 7.3 (Σ-typed variable) Let Σ be a signature and let X be a set of symbols such that each $x \in X$ is assigned a domain symbol, $\underline{A}$, of Σ. Then x is a Σ-*variable of type* $\underline{A}$.

The constants, functions, and variables determine the *terms* of a signature.

Definition 7.4 (terms) Let Σ be a signature and let X be a set of Σ-variables. Then the *terms over* Σ *and* X are defined as follows:

- each constant symbol of type $\underline{A}$ is a term of type $\underline{A}$;
- each variable of type $\underline{A}$ is a term of type $\underline{A}$;
- given a function symbol $\underline{f} : \underline{A_1} \times \cdots \times \underline{A_n} \to \underline{A_{n+1}}$ and term t_1 of type $\underline{A_1}$, term t_2 of type $\underline{A_2}$ et cetera. Then the symbol sequence $\underline{f}(t_1, \ldots, t_n)$ is a term of type $\underline{A_{n+1}}$.

For example, with signature *bakery* and with x a variable of type *pastries*, $\underline{next}(\underline{des}(x))$ is a term of type *pastries* over the signature *bakery*.

7.4 What a term denotes: meaning of terms

A term u of type M over a structure S, and a set X of variables together with an *assignment* for X, specifies an element of M.

Definition 7.5 (assignment of variables) Let Σ be a signature, let X be a set of Σ-variables, and let S be a Σ-structure. An *assignment of X in S* assigns each $x \in X$ of type $\underline{A}$ an element $ass(x) \in A_S$.

The meaning of a term in a structure for a given assignment of the involved variables is now defined as follows.

Definition 7.6 (meaning of terms)   Let Σ be a signature, let X be a set of Σ-variables, let S be a Σ-structure, and let *ass* be an assignment of X. For each term u over Σ, the *meaning of u in S*, written: $[u]_S$, is an element of a domain of S, defined as follows:

- $[\underline{c}]_S = c_S$ for each constant symbol $\underline{c}$;
- $[x]_S = ass(x)$ for each variable x;
- $[\underline{f}(t_1,\ldots,t_n)]_S = f_S([t_1]_S,\ldots,[t_n]_S)$ for each term of the form $\underline{f}(t_1,\ldots,t_n)$.

For example, with signature *bakery*, variable x of type *pastries*, term $\underline{next}(\underline{des}(x))$, the structure S = *three pastries*, and $ass(x) = pie$, holds:

$$
\begin{aligned}
[\underline{next}(\underline{des}(x))]_S &= \\
next_S([\underline{des}(x)]_S) &= \\
next_S(des_S(\underline{x})) &= \\
next_S(des_S(ass(x))) &= \\
next_S(des_S(pie)) &= \\
next(\text{``pie''}) &= \text{``bread''}.
\end{aligned}
\tag{7.1}
$$

Likewise, with structure S = *five pastries* holds:

$$
[\underline{next}(\underline{des}(x))]_S = \text{``rol''}.
\tag{7.2}
$$

Predicate logic

Predicate logic poses the fundamental question as to what extent elements of the real and imagined world can be symbolically represented, and how properties of those elements can be deduced from those representations. Typical introductory texts on predicate logic, e.g. [20], usually start with a signature Σ – often called *syntax* – and ask for properties of all or of specific classes of Σ-structures; Σ-structures are frequently called "models" of Σ. Other texts on logic focus on the intuitive justification of rules to deduce properties of Σ-structures [89].

Since the 1970s, informatics has based the specification of items and data on signatures and their interpretations. Compared to predicate logic, signatures may be very large, but interpretations come with finite or at most countably infinite domains, and standard data types are imported. Typical examples include *VDM, Z*, and many others.

7.5 How to specify static aspects

To capture a system of the real or an imagined world, one starts out by covering static, unvarying aspects of a system as a structure S, according to Section 6.2, page 69.

Luckily, each structure can be assigned a signature Σ and hence be turned into a Σ-structure, as defined in Section 7.2, page 74. A signature Σ for S then provides the – in general infinite – set of Σ-terms, as well as the assignment of a meaning for each term. Not all aspects of a system can be represented by terms. But the well-organized nature of structures and the inductive build-up of terms and their meaning in corresponding signatures provide the best means for this enterprise.

Modeling is frequently intended not to describe just *one* system, but a set of similar systems, as defined in Section 7.2, page 74. For example, as mentioned above, the *bakery* signature of Figure 7.1, page 74, covers both the structure *three pastries* of Figure 6.1, page 70, and the structure *five pastries* of Figure 6.2, page 71: both these structures are structures, among many others, of the signature *bakery*.

In general, the set of all Σ-structures of a signature Σ is huge, and one is only interested in a subset of those structures, with particular properties. Such properties are characterized by means of *requirements* on Σ-structures.

As an informal characterization, a *static specification* over a signature Σ consists of Σ and a set of requirements. For a signature Σ, a Σ-structure S is a *model* of a static specification over Σ, if its requirements hold in S. For example, the "intended" structures of the signature *bakery* assume that the function *des* reverses the function *item* – and vice versa. This is expressed by the two requirements:

$$\text{for all } x \in pastries : item(des(x)) = x; \tag{7.3}$$

$$\text{for all } y \in descriptions : des(item(y)) = y. \tag{7.4}$$

For a static specification of the *three pastries structure* of Figure 6.1, page 70, we add the requirement

$$\text{for all } x \in pastries : \underline{item}(\underline{next}(\underline{next}(\underline{next}(\underline{des}(x))))) = x. \tag{7.5}$$

The *five pastries* structure of Figure 6.2, page 71, does not meet this specification. For example, with the assignment $ass(x) =_{\text{def}} bread$ holds:

$$\underline{item}(\underline{next}(\underline{next}(\underline{next}(\underline{des}(x)))))_{\text{five pastries}} = rol. \tag{7.6}$$

In the HERAKLIT framework, each module comes with a signature Σ. In this capacity, the module and its signature model a *system schema*. In addition, one or a small number of "intended" Σ-structures are identified. The more requirements are formulated, describing properties of the intended structure(s) S, the more can formally be proven about S.

Later on, we will see that each HERAKLIT module comes with its own structure or signature. Therefore, the composition of modules requires the composition of structures or signatures which can be easily achieved by the building the union of the structures or signatures under consideration. A more detailed discussion will be given in future publications.

Related work to Part III

Historically, *universal algebra* is the field of mathematics that studies properties of sets of algebraic structures [32]. Such a structure S usually includes a small number of sets, constants, functions, and relations (predicates). A corresponding collection of symbols, a *signature* Σ, includes a symbol for each of these sets, constants, functions, and relations. In general, many structures correspond to a signature Σ. The symbols of Σ give rise to *terms* and *formulas*, which predicate logic uses in order to characterize classes of structures and to study specific properties. Typed structures are the backbone of the *Bourbaki* approach to a systematic presentation of mathematics [14]. In the meantime, this approach prevailed not only in mathematics, but also in empirical sciences. [90] exemplifies this in physics, philosophy, psychology, computer science, economics, and semiotics.

Informatics took up these concepts, under the name of *algebraic specification*, to handle data and items of any kind [82]. However, in informatics, signatures and corresponding structures are often much bigger than in classical universal algebra. Specification languages to manage algebraic specifications include *Alloy, B, Common Algebraic Specification Language* (CASL), *CafeOBJ, Rigorous Approach to Industrial Software Engineering (RAISE), VDM, Z*, and many others.

The first proposals to cope with individual items in Petri nets were based on universal algebra, represented as moving from propositional logic to first-order logic [31]: a Petri net place is a predicate that applies to a varying set of items. HERAKLIT follows this line of handling data and items, enhancing nets by:

- conceiving a local state of a run as a predicate that applies to an item;
- explicitly distinguishing schemata and their instantiations;
- generating a set of tokens from a (fitting) term by means of the *elm*-notation which will be introduced in Section 9.7, pages 102f.

Some process algebras, most notably *mCRL2* [33], specifically integrate process algebra with data structures.

Business informatics typically uses the *entity-relationship model (ERM)* to model static aspects [17]. ERM exists in hundreds of different variants [66], all addressing particular application needs. For example, SAP uses *structured ERM* for static modeling [91].

Part IV
Consolidating the three pillars:
HERAKLIT modules

The above three parts of this book separately consider the three pillars of HER-
AKLIT, i.e. *architecture*, *dynamics*, and *statics* of HERAKLIT models. The concepts
of each single pillar alone suffice to model particular kinds and aspects of systems;
for example, the overall structure of a system can be covered by the concepts of
Part I. A system with distributed control, but without data, can be dealt with with
the concepts of Part II. A specification that deals only with real-world items and
data structures and their symbolic representation may only use Part III. However, in
general, a comprehensive model of a really large system must represent architecture,
behavior, and real- or imagined-world items and data, in an integrated form. It turns
out that the concepts of the three HERAKLIT pillars together essentially leverage the
expressive power of the single pillars, and add up to a particularly powerful formal
framework. This integrated view of the three HERAKLIT pillars is the topic of this
Part IV.

Chapter 8
Items and behavior:
general system modules and their runs

The central and most comprehensive notion of HERAKLIT models is *general system modules*. Such a module frequently characterizes a set of similarly formed elementary system modules. This characterization exploits the observation that a state of an elementary system module is frequently a predicate *p* together with an argument *c* for *p*, as defined in Section 6.1, page 67. In this case it is useful not to construct a single place for each argument *c* of *p*, but to represent the predicate *p* itself as *one* place, with the corresponding arguments *c* as inscriptions. Furthermore, all occurrences of events that affect the same predicates are modeled by *one* transition. The transition's name is parameterized by variables that can be assigned respective elements. Each assignment of the variables in those terms characterizes one of the original events. Hence, a *general system module* is a net module, as introduced in Section 4.2, page 43. Each place is a *predicate*, and each transition represents a set of *events*.

8.1 Motivating example

We return to our running example of a bakery, and refine the model's steps. For example, the step *bake* in Figure 4.2, page 44, in Section 4.3 is now refined into three steps: *bake bread*, *bake cake*, and *bake pie*, as shown in Figure 8.1, page 86. There, the structure *bake* shows the underlying domains et cetera, as a variant of the structure *three pastries* of Figure 6.1, page 70, in Section 6.2. Each of the three steps comes with two states. Each of the states is an instantiation of one of the two predicates, *ready to bake* and *on counter*. Each instantiation calls one of the pastries, *bread*, *cake*, or *pie*, as well as its description. This example illustrates a general observation: whenever real-world items or data are involved in a step, the propositions that constitute the states of the step are frequently instantiated predicates as discussed in Section 6.1, page 67.

As a second example, the *supply to aide* step of Figure 4.2, page 44, in Section 4.3 is likewise parameterized, additionally using the predicates *on counter*, *recent supply*,

P. Fettke, W. Reisig, *Understanding the Digital World*,
https://doi.org/10.1007/978-3-031-61898-7_8

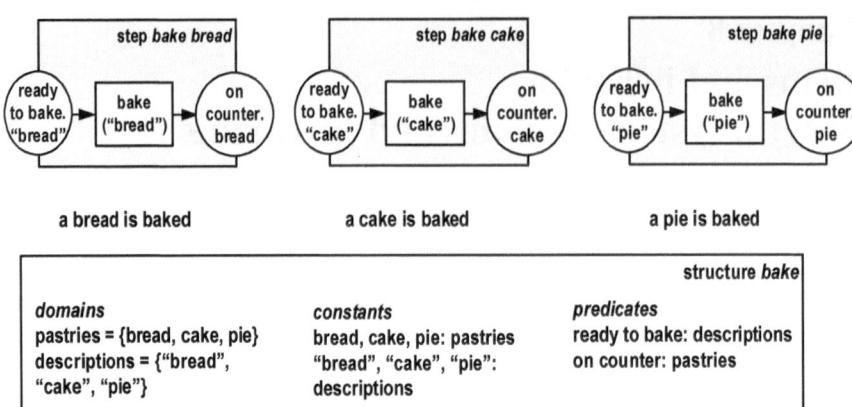

Fig. 8.1 Local states are instantiated predicates.

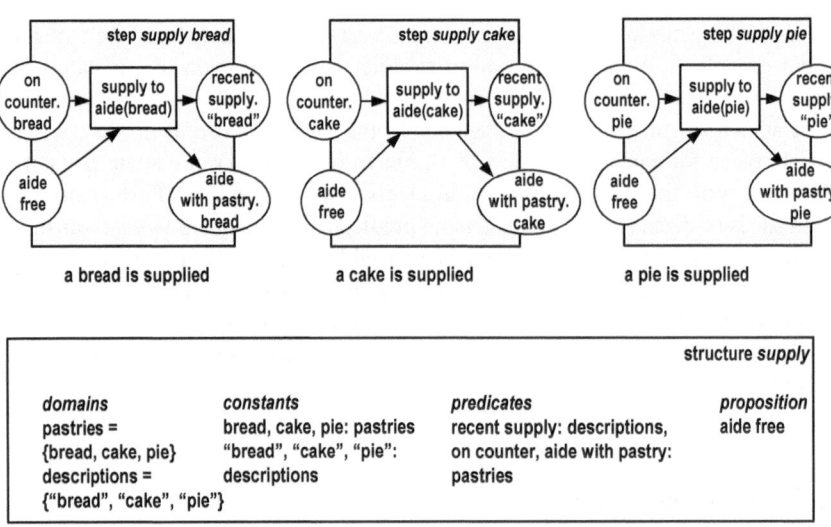

Fig. 8.2 Different pastries are supplied to the aide.

and *aide with pastry*, as well as the proposition *aide free*. Figure 8.2 shows these modules.

Finally, Figure 8.3, page 87, shows steps that select the respective next step for each recently supplied pastry.

Corresponding parameterized versions of the *move to shop* and the *sell* modules of Figure 4.2, page 44, are now obvious.

Altogether, instances of these steps can now be composed into the run in Figure 8.4, page 87, representing the lifeline of a bread, from being baked to being sold. This run is extended by corresponding lifelines for cake and pie, as shown in Figure 8.5, page 87.

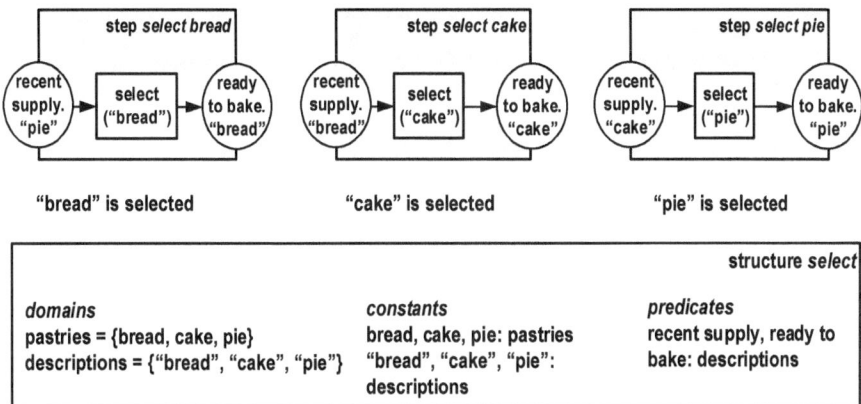

Fig. 8.3 The pastry to be baked next is selected in terms of descriptions.

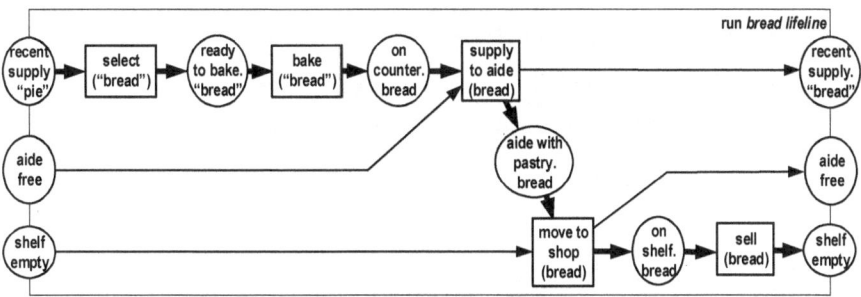

Fig. 8.4 Lifeline of bread (bold-faced arrows): First *"bread"* is selected, then a bread is baked, supplied, moved to the shop, and sold. Algebraically this is written: *bread lifeline* =$_{\text{def}}$ *select("bread") • bake("bread") • supply to aide(bread) • move to shop(bread) • sell(bread)*.

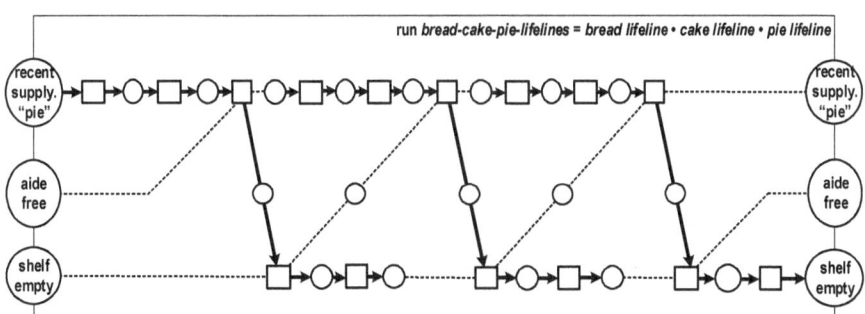

Fig. 8.5 Three lifelines: *bread lifeline • cake lifeline • pie lifeline*. To increase readability, inscriptions of inner elements are skipped. Lifelines are bold faced, other causal dependencies are dotted.

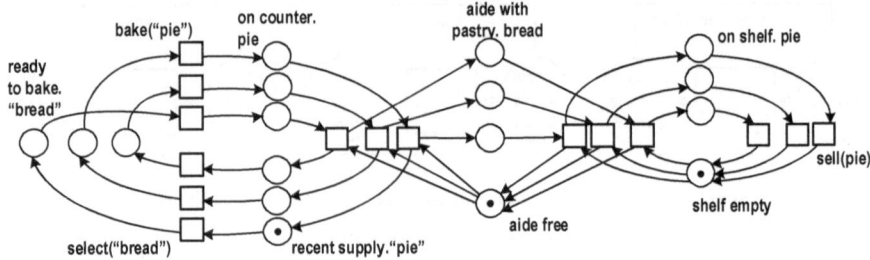

Fig. 8.6 Elementary system for three kinds of pastry. For the sake of readability, inscriptions on transitions are omitted, and inscriptions on places are bundled.

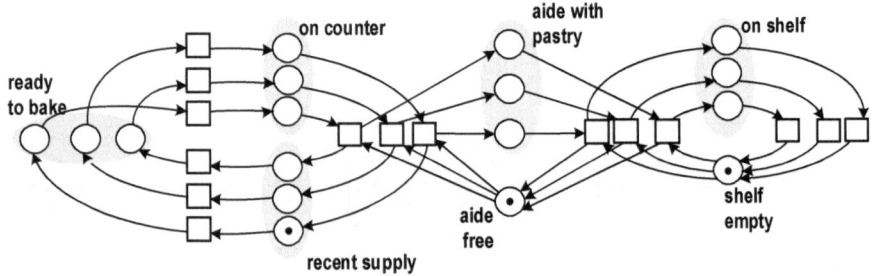

Fig. 8.7 From propositions to predicates.

8.2 From propositions to predicates

Figure 5.2, page 54, in Section 5.1 represents all manifold iterations of the run of the bakery system as one HERAKLIT system module. A corresponding system module is shown in Figure 8.6 for the case of three different pastries in the cyclic order bread, cake, pie, bread, et cetera.

Figure 8.6 is at the brink of readability. A similar pictorial representation for the case of, say, 10, 100, or even more kinds of pastry is not manageable. Other means to represent systems with many, similarly shaped modules are inevitable. Here, the notions and concepts of Part III come in.

Conceptually, we exploit the above observation that in system models such as in Figure 8.6, a state is frequently an instantiation of a predicate. Figure 8.7 sketches this observation. In Figure 8.8a, page 89, the constants *"bread"*, *"cake"*, and *"pie"* are instantiations of the predicate *recent supply*. The same constants are also instantiations of the predicate *ready to bake*. Figure 8.8b now shows the decisive step: both predicates are represented themselves, as two net places, labeled *recent supply* and *ready to bake*. Figure 8.8a shows that the state *recent supply. "pie"* is actually reached. In Figure 8.8b, this is represented by the inscription *"pie"* of the predicate *recent supply*. Figure 8.8c shows the underlying structure *select*.

The function *next* describes that the three pastries are baked in cyclic order: Baking (and supplying) pie is followed by baking bread. Baking (and supplying)

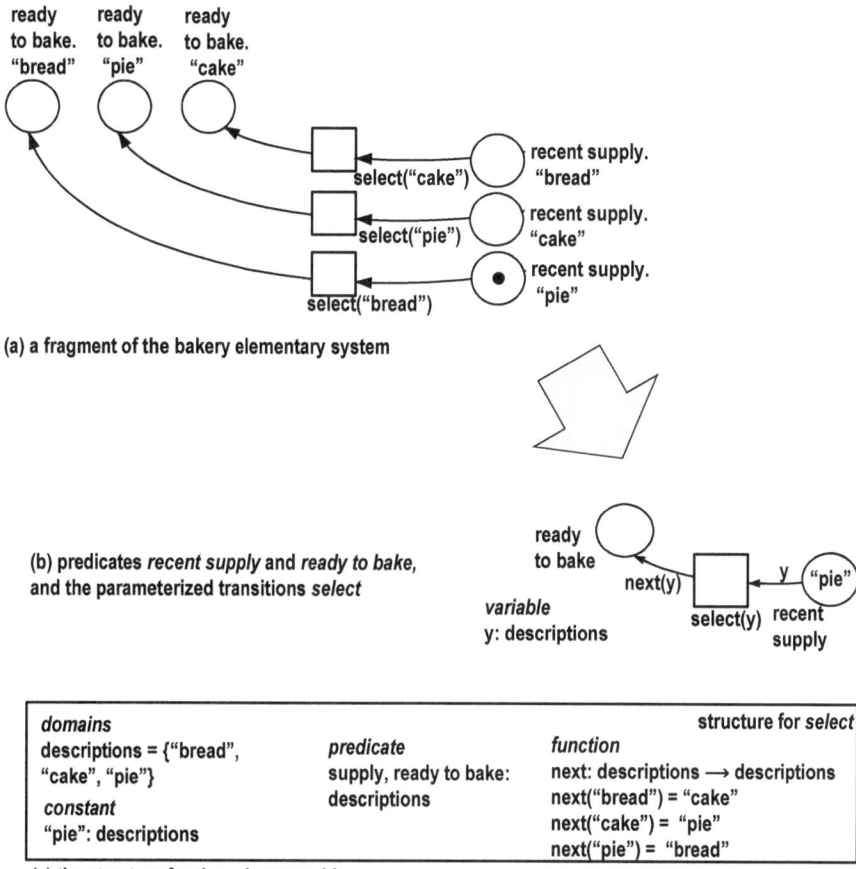

(a) a fragment of the bakery elementary system

(b) predicates *recent supply* and *ready to bake*, and the parameterized transitions *select*

(c) the structure for the select transition

Fig. 8.8 From propositions to predicates.

bread is followed by baking cake. And baking (and supplying) cake is followed by baking pie. Notice that this is described on the level of descriptions, not on the pastries themselves.

Generally, actually reached states are represented by the corresponding constants written inside the place. Correspondingly, the events – the event *select* in Figure 8.8 – are parameterized, i.e. adopt the character of functions, with variables as parameters, varying over given domains.

As a second example, Figure 8.9, page 90, represents the step from single instantiations of the predicates *ready to bake* and *on counter* to the predicates themselves.

Figure 8.10, page 91, repeats the above move from propositions to predicates for the case of the three *supply to aide* events and the adjacent states. The function *des* assigns each pastry its description. The place labeled *aide free* represents a proposition.

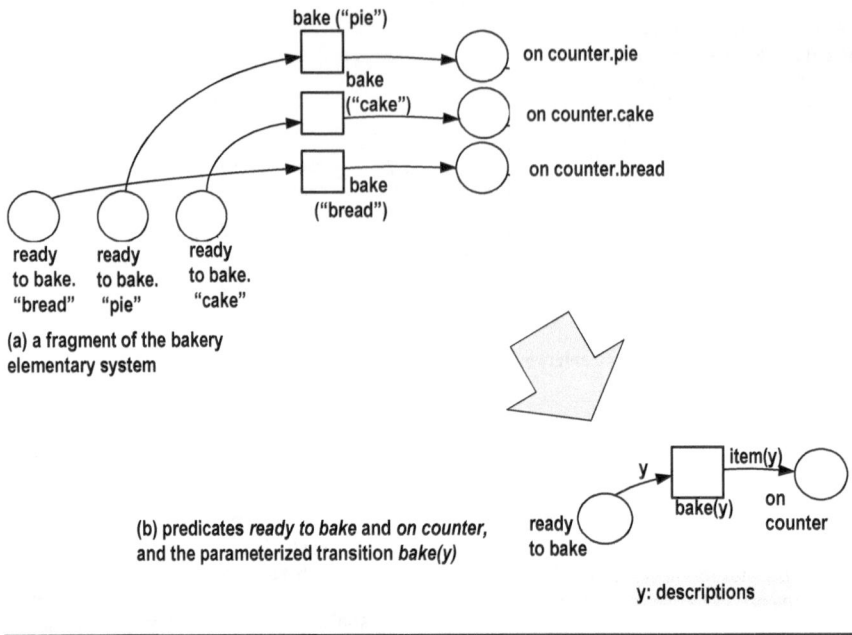

(a) a fragment of the bakery
elementary system

(b) predicates *ready to bake* and *on counter*,
and the parameterized transition *bake(y)*

(c) the structure for the bake transition

Fig. 8.9 From propositions to predicates.

Combining the predicate-based versions of the *select*, the *bake*, and the *supply* fragments yields the *system bakehouse* module in as in Figure 8.11 (top left), page 92. Accordingly, the *system shop* module of Figure 8.11 (top right) is obtained from Figure 8.6, page 88. The entire bakery is now obtained as the composition of the bakehouse module with the shop module, as in Figure 8.12, page 93.

This *general system module* in fact perfectly represents the elementary system of Figure 8.6, page 88. This system, in turn, describes the runs of Figures 8.4 and 8.5, page 87, and their extensions. These runs can directly be derived from the bakery system in an intuitively obvious way.

8.3 General system modules

We are now prepared to precisely describe general system modules such as the *bakery with pastries* module of Figure 8.12, page 93: each module comes with a

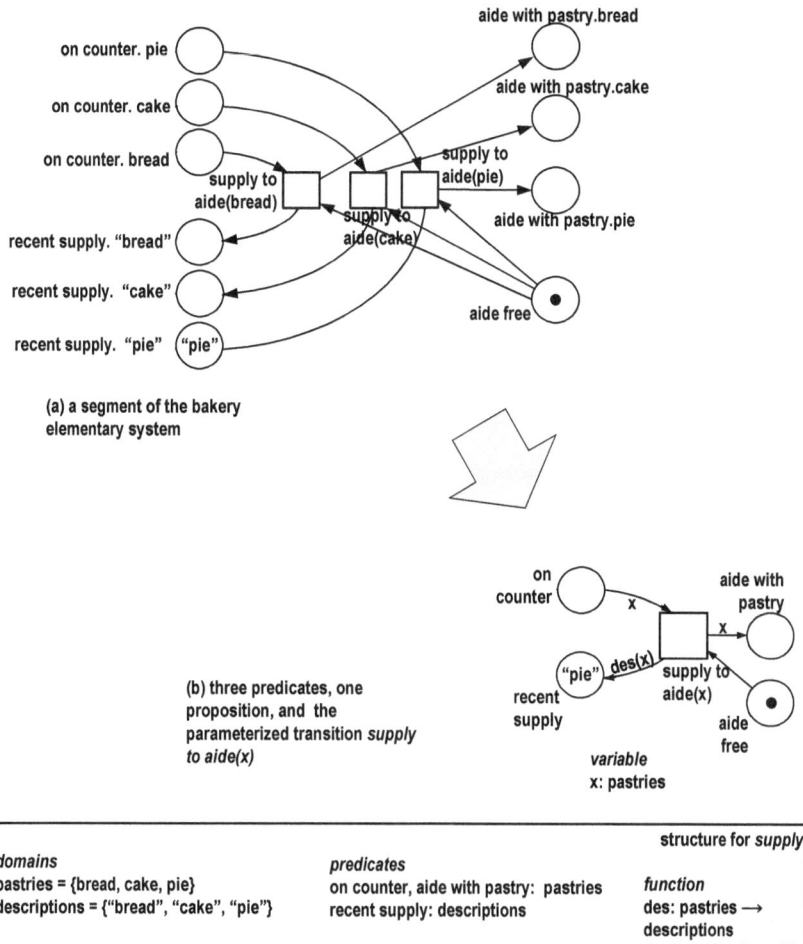

Fig. 8.10 Another step from propositions to predicates.

Σ-structure, S. The domains, constants, and functions of S describe static aspects. Each predicate is a place of the module's net, each edge is inscribed with a term, and each transition, together with its adjacent arcs, describes one of the system's steps.

Definition 8.1 (general system module) Let Σ be a signature, let X be a set of Σ-variables, and let S be a Σ-structure. Let N be a net module where:

- each place p of N is a predicate of S;
- each edge (p, t) or (t, p) of N is inscribed with a term u over Σ and X, where the type of u coincides with the type of p.

Then *N* is a *general system module over S and Σ with variables X*.

Figures 8.11 and 8.12, page 92, show examples. As in the case of elementary system modules, an initially assumed state is frequently distinguished and represented by constants in the corresponding place symbols.

In the above examples, it is assumed that initially, *recent supply* has been *"pie"*, and that both propositions *aide free* and *shelf empty* are reached. Elementary system modules can be conceived as general system modules over the trivial structure, as defined in Section 6.2, page 69:

Observation and graphical convention:

1. Each elementary system module is a general system module. Its places are not proper predicates, but only propositions.
2. In graphical representations of elementary system modules, we skip the dot inscription on edges.

Composing the bakery from step modules

One may wonder whether step modules as defined in Section 4.3, page 43, can be used, in an intuitive way, to compose system modules such as the system *bakehouse* of Figure 8.11, page 92, or the system *bakery with pastries* of Figure 8.12. To achieve this, Figure 8.13, page 93, slightly re-draws the

---

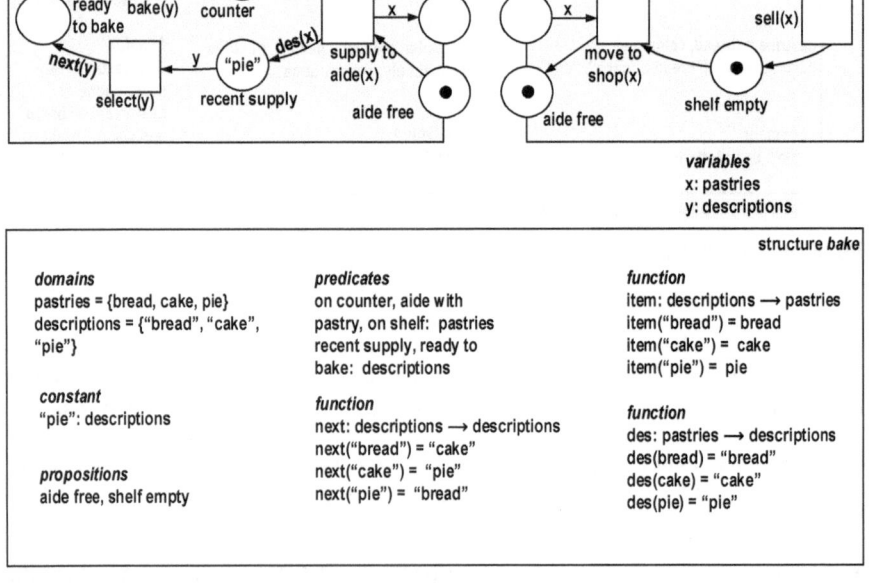

Fig. 8.11 Predicate based producer and consumer.

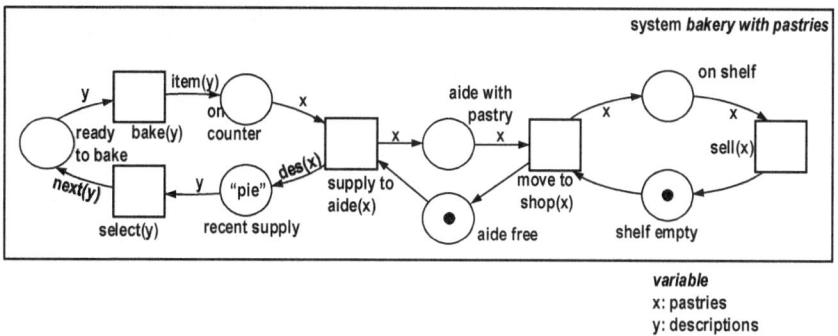

Fig. 8.12 *bakery with pastries* =_{def} *bakehouse • shop* with structure from Figure 6.1, page 70.

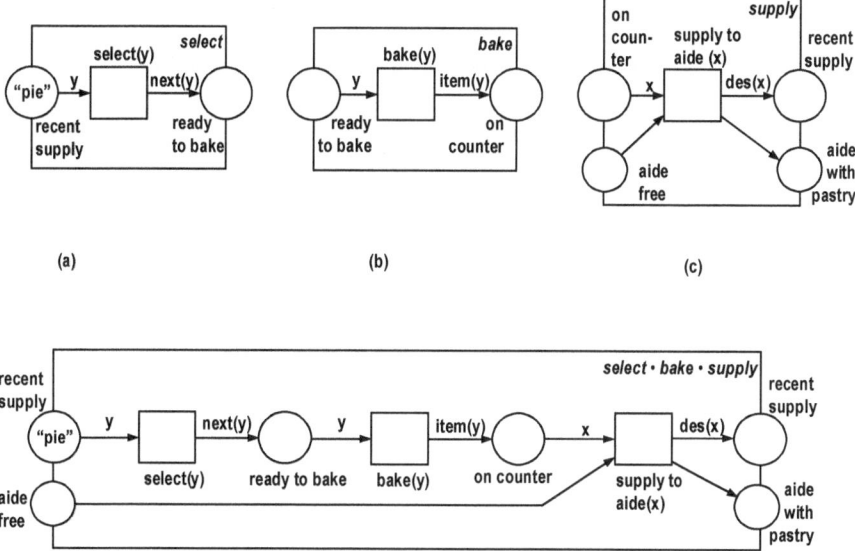

Fig. 8.13 Pastry steps and their composition.

steps as in Figure 8.3, page 87, and turns them into step modules. This is shown in Figure 8.13a-c. Their composition, as in Figure 8.13d, differs from the *bakehouse* module of Figure 8.11. However, this can easily be repaired, merging the *recent supply* labeled places in the left and right interfaces, and moving it to the interior of the module.

8.4 Steps and runs of general system modules

For a transition t of a general system module N, each assignment of the involved variables generates a *step*, as defined in the previous Section 8.3. Intuitively, for a given assignment of the variables, the term u of an edge (p,t) of N fixes an item $[u]$ and hence a state $p.[u]$. This state is assumed to be presently reached. The step assumes "flows" from p towards t. Correspondingly, the assignment fixes an item $[u]$ for the term u of an edge (t,p), that "flows" from t towards p, hence the state $p.[u]$ will be reached.

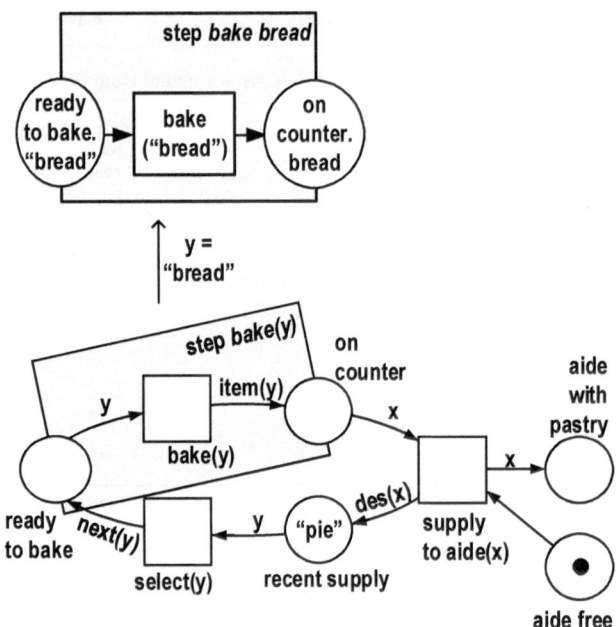

Fig. 8.14 The step *bake("bread")* of the *bakehouse* module.

Figure 8.14 shows an example of a step of the *bakehouse* module. In the formal framework one would define steps by these technicalities.

Definition 8.2 (step of general system module) Let Σ be a signature, let S be a Σ-structure, and let X be a set of Σ-variables. Let N be a general system module over S with variables X, and let t be a transition of N. Let ass be an assignment of X. For each term u at arcs adjacent to t, let $[u]$ denote the meaning of u as defined in Definition 7.6, page 76, in Section 7.4. Furthermore, let D be a step module with e its transition. Assume for each place p of N

1. (p, t) is an edge of N if and only if for the term u of (p, t), the left interface $^*D$ contains a $p.[u]$-labeled place q, and (q, e) is an edge of D;
2. (t, p) is an edge of N if and only if for the term u of (t, p), the right interface D^* contains a $p.[u]$-labeled place q, and (e, q) is an edge of D.

Then D is a *step* of N, generated by t and ass.

For example, with the assignment $y = $ "*bread*", the $bake(y)$ transition of Figure 8.11, page 92, and of Figure 8.12, page 93, generates the *bake bread* step of Figure 8.1, page 86. Figure 8.14 sketches this. Likewise, $y = $ "*cake*" and $y = $ "*pie*" generate the *bake cake* and *bake pie* steps. And the three steps of Figure 8.2, page 86, are steps of the general system module *bakery with pastries*, generated by the transition *supply to aide* and the three different assignments of the variable x of type *pastry*.

Based on the notion of run modules as defined in Section 4.4, page 44, and in analogy to the definition of runs of elementary system nets, runs on general system nets are defined as follows.

Definition 8.3 (run of a general system module) Let S be a structure. Let N be a general system module over S, and let D be a run module such that each step module of D is a step of N. Then D is a *run of* N.

The run *bread lifeline* of Figures 8.4 and 8.5, pages 87 and 87, is in fact a run of the general system module *bakery with pastries* in Figure 8.12, page 93.

A crucial property of runs is their composability.

Theorem 8.1 *The composition of runs of a general system module N is again a run of N.*

Proof. To prove this theorem, notice that in Section 5.2, page 54, we proved the theorem for elementary system modules already. □

The box on predicate logic at the end of Section 7.4, page 76, emphasizes that an algebraic specification essentially characterizes static system aspects. Nevertheless, a couple of algebraic specification techniques suggest – often rudimentary – means to represent also dynamic aspects: A global state is represented as a Σ-structure, and a run of a system is defined as a sequence $S_1, S_2, \ldots$ of such states S_i. The problem then is to define the *successor* function, *suc*, that assigns each state S_i its successor state S_{i+1}. Particularly

interesting is the approach of an *abstract state machine* (ASM) [34], which
defines successor states by means of assignment statements of the form:

$$f(a) =_{\text{def}} g(b), \tag{8.1}$$

where $f(a)$ and $g(b)$ are terms of the underlying signature Σ. In this setting,
HERAKLIT can be conceived as a variant of this idea: Σ-terms retain their
values, but predicates update the set of elements to which they apply.

8.5 Commenced system modules

To conceive a system model, it is intuitively helpful to indicate a set of "*initially
assumed*" states, and to concentrate on runs R with the left interface, $^*R$, representing
the initially assumed states.

Definition 8.4 (commenced general system module) Let N be a general system
module over a structure S. Let *init* be a mapping that assigns each place p of S a set
of constants of the type of p. Then N is *commenced*.

Definition 8.5 (run of commenced general system module) Let S be a structure
and let N be a commenced general system module over S. Let D be a run of N, such
that for each place p of N and each constant c of the type of p holds: if $^*D$ contains
a (p, c)-labeled place, then c is in $init(p)$. Then D is *commenced*.

Intuitively formulated, a commenced run of a commenced system module N starts
with the states prescribed by the initially assumed states of N. In fact, all elementary
and all general systems considered so far – except the elementary system module in
Figure 5.4, page 57 – have been commenced, and all runs of those systems considered
so far have commenced as well.

Continuing the discussion of conflicting situations in Section 5.3, page 56, it is
interesting to observe that absence of conflicting situations implies that only one
maximal commenced run exists.

Definition 8.6 (deterministic commenced general system module) Let S be a
structure. Let N be a commenced general system module over S.

1. A run D of N is *maximal* if and only if

   - for each place p of N and each constant c holds: if c is in $init(p)$ then $^*D$
     contains a (p, c)-labeled place, and
   - D cannot be expanded, i.e. for no step E, $D \bullet E$ is a commenced run of N.

2. N is *deterministic*, if N has only one maximal commenced run.

In fact, all elementary and all general systems considered so far are deterministic,
except the elementary system module in Figure 5.4, page 57.

Chapter 9
Variants of the bakery:
flexible modeling with HERAKLIT

To demonstrate the intuitive perspicuity of HERAKLIT models and the convenience of modeling with HERAKLIT, in this section we emphasize some particularly useful features of general system models, and slightly extend their expressivity. All features are exemplified by variants of the bakery. Some modules are deterministic, as defined in Section 8.5, page 96, others are not. The forthcoming examples support the claim that HERAKLIT provides exactly what is intuitively required. In particular, sets and multi-sets turn out to be generally useful.

9.1 Specialized shop assistants: terms without variables

So far, an edge of a general system module is inscribed with a variable such as "x", or with a term of the form $f(x)$, with "f" a function symbol. As defined in Definition 7.4, page 75, of Section 7.3 technically, any term, including the constants, may occur as an edge inscription of a general system module. Figure 9.1 shows a variant of the bakery: each kind of pastries is sold in a separate area of the shop, or by a specialized shop assistant; so, three different *sell* events are distinguished. For each specific pastry on the shelf there exists a unique event that describes its sale. So, despite three edges starting *on shelf*, any pastry on place *on shelf* solicits exactly one of the three *sell* events to occur. In addition, the amount of sold bread is counted. Altogether, the system *bakery with pastry as constants* is deterministic. Figure 9.2 shows an initial part of the run of this system. This run extends the run of Figure 8.5, page 87, by the counter for sold bread.

9.2 The baker decides: alternative runs

In the *bakery with pastries* model of Figure 8.12, page 93, the baker bakes different pastries in a cyclic order, defined by the *next* function of the underlying structure.

© The Author(s), under exclusive license to Springer Nature Switzerland AG 2024
P. Fettke, W. Reisig, *Understanding the Digital World*,
https://doi.org/10.1007/978-3-031-61898-7_9

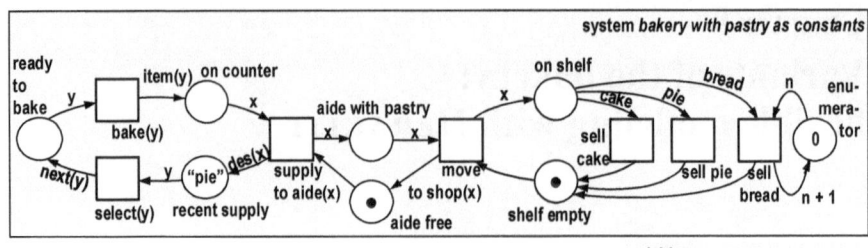

variables
x: pastries y: descriptions n: N

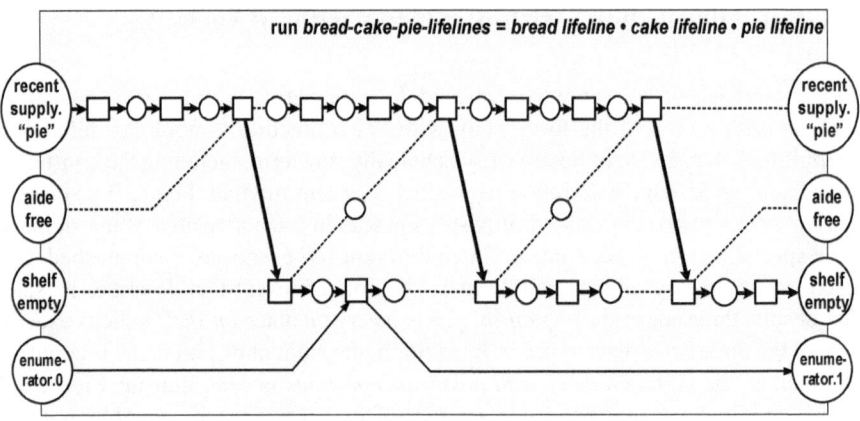

Fig. 9.1 Each kind of pastry is sold by a specialized activity. Sold bread is counted.

Fig. 9.2 Selling and counting a bread.

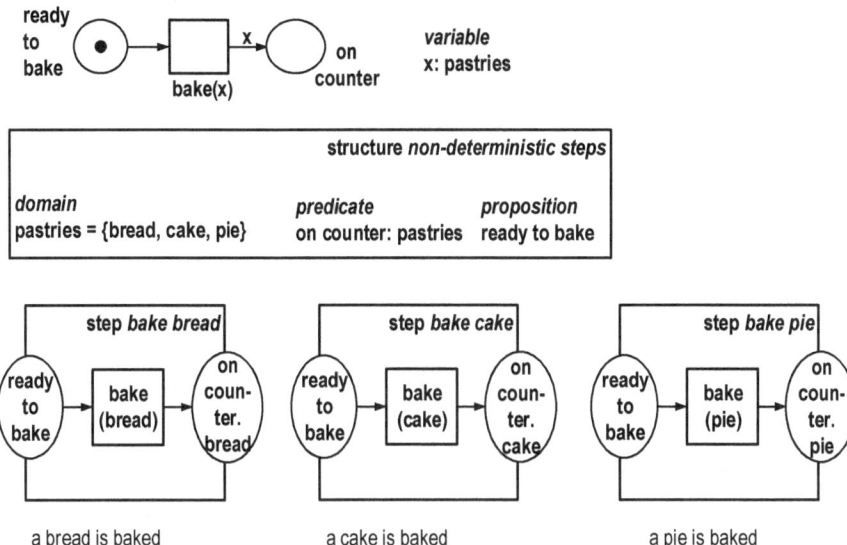

Fig. 9.3 A fragment of the forthcoming module *non-determinism*, and three different steps of *bake(x)*, all starting with the local state *ready to bake*.

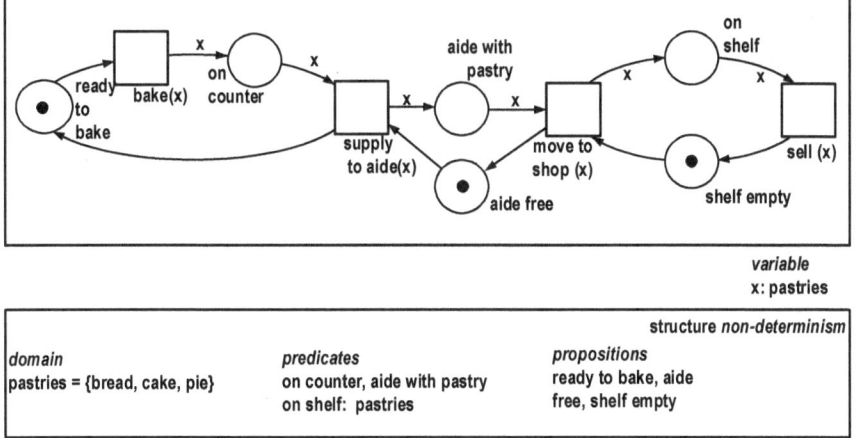

Fig. 9.4 Three alternative assignments of the variable *x*.

Now, we want to model the case where the baker autonomously chooses the pastry he intends to bake next. Figure 9.3 shows a transition, *bake(x)*, that generates three steps, all starting with the state *ready to bake*, and ending with different instantiations of the *on counter* predicate. These steps are the first steps of different runs of the module *non-determinism* of Figure 9.4. Each time a run reaches the state *ready to bake*, again three alternative continuations appear.

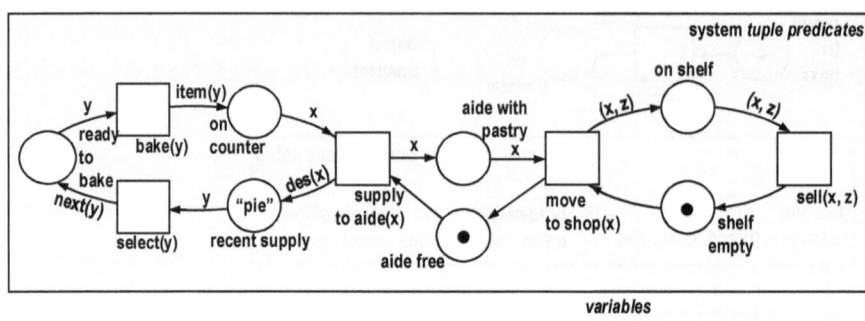

Fig. 9.5 Each pastry on the shelf is decorated with a price.

9.3 Pastry with price tags: predicates with tuple-domains

In all cases considered so far, a predicate is just a domain, and hence, a step updates a single item of a single predicate of a given structure. Frequently, a predicate p is a subset of a cartesian product $M_1 \times \cdots \times M_n$ of domains M_i, and hence, a step updates a tuple $(m_1, \ldots, m_n)$. Figure 9.5 shows a typical example: As discussed in the previous Section 9.2, page 97, in system *non-determinism* of Figure 9.4, the variable x can be assigned any of the three pastries. Now, in addition, the variable z varies over the prices, i.e. all infinitely many natural numbers. Intuitively formulated, the predicate *on shelf* applies to tuples consisting of a pastry, x, and a price, z. Choice of the price is not specified.

9.4 Fixed sets of items and shelves: commenced states

So far, the bakery model assumes one shelf with space for just one pastry. Figure 9.6 models the case of *six* shelves. Technically, *empty shelves* is a predicate that initially applies to six shelves, a, b, c, d, e, and f. A pastry on a shelf is now modeled as a triple: the pastry, its price, and its shelf. In case of more than one empty shelf, the

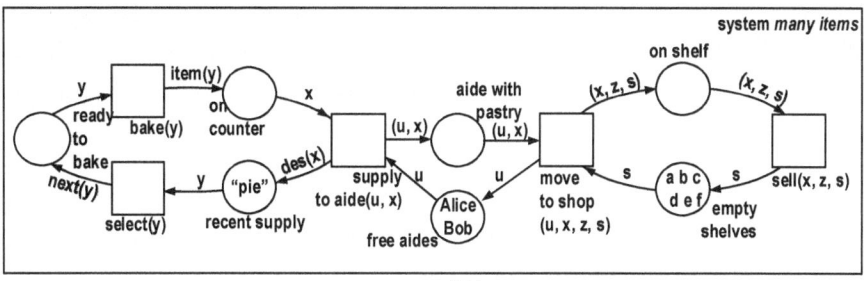

Fig. 9.6 Predicates *free aides* and *empty shelves* with many initial instantiations.

event *move to shop* may occur non-deterministically by assigning the variable s any of the empty shelves.

Additionally, we assume *two* aides, Alice and Bob. For each pastry, the *supply to aide* transition determines which of the two aides accepts the pastry and moves it to the shop. Therefore, the *aide with pastry* predicate is applicable to tuples *(aide, pastry)*, and the *supply to aide* transition has two variables, *u* and *x*. The variable *u* varies over the two aides, Alice and Bob. Of course, this system is non-deterministic.

Likewise, the *on shelf* predicate applies to triples *(pastry, price, shelf)*. None of the incoming edges of the *move to shop* transition carries the variable "*z*"; hence, the variable *z* can be assigned any price, i.e. any natural number, as in Figure 9.5, page 100. Generally, an assignment to all variables in the inscriptions of the surrounding edges of a transition *t* defines a step of *t*. As an example, the run in Figure 9.7, page 102, is a run of the system in Figure 9.6, page 101. The step of Figure 8.14, page 94, occurs in this run. The run is not commenced: in its *start*, there are no places with labels *free aides.Alice*, *empty shelves.d*, *empty shelves.e*, or *empty shelves.f*.

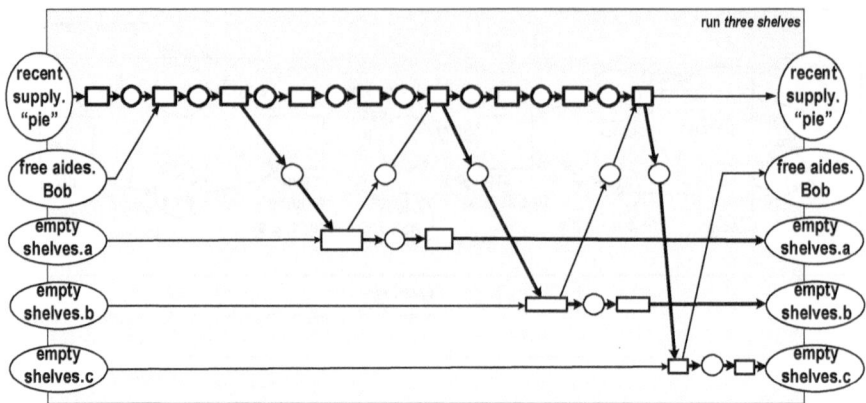

Fig. 9.7 Shelves a, b, and c are used for a *rol*, a *biscuit*, and a *bread*. The other three shelves are not in use in this run.

9.5 Selling bags of pastry: sets of updates

So far, the bakery sells single pastries. Now we intend to model the case where they sell *bags* of pastry. Each bag contains a *bread*, a *cake*, and a *pie*. Figure 9.8 shows this case. In technical terms, the edge from place *aide with pastry* to the transition *move to shop* is inscribed with three constants: *bread*, *cake*, and *pie*. With those three items in the place *aide with pastry*, the transition *move to shop* produces *one* item, a set containing the *bread*, *cake*, and *pie* on the place *on shelf*. This set, representing a bag, is then sold.

9.6 All breads are alike: multi-sets

Pastry of one sort needs no further distinction: two items of bread are alike in every context. The module *bags of pastries* in Figure 9.9 packs two alike pastries in one bag and stores the bag on one of two shelves. Here we assume the mathematical notion of multi-sets; finite multi-sets are also called *bags*, where each item may occur in many indistinguishable copies. Multi-sets are usually written in square brackets. The place "empty shelves" is no longer a proposition. It contains up to two black dots, representing a predicate that applies to many "black dots".

9.7 Most liberal baker

In the system module of Figure 9.4, page 99, in Section 9.2, the baker freely decides each time which pastry he intends to bake. Now, intuitively formulated, each time

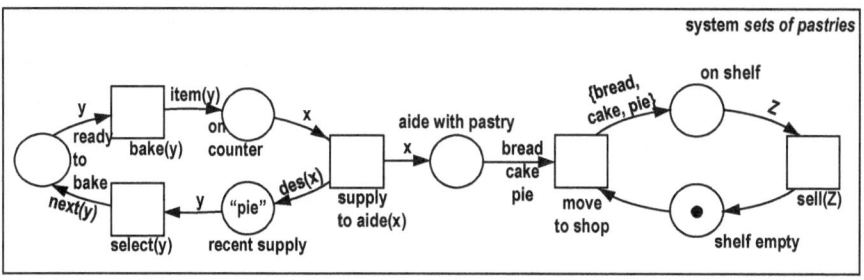

variables
x: pastries y: descriptions Z: set of pastries

structure sets of pastries		
domains	predicates	function
pastries = {bread, cake, pie}	on counter, aide with pastry:	next: descriptions ⟶
descriptions = {"bread", "cake",	pastries	descriptions
"pie"}	recent supply, ready to bake:	next("bread") = "cake"
	descriptions	next("cake") = "pie"
constants	on shelf: set of pastries	next("pie") = "bread"
"pie": descriptions		
bread, cake, pie: pastries	function	function
	item: descriptions ⟶ pastries	des: pastries ⟶ descriptions
proposition	item("bread") = bread	des(bread) = "bread"
shelf empty	item("cake") = cake	des(cake) = "cake"
	item("pie") = pie	des(pie) = "pie"

Fig. 9.8 Selling bags of pastries.

the baker bakes a set of pastries on a tray and places the tray on the counter. The number and kind of pastries on the tray may vary and is left unspecified. The baker may bake a different kind and number of pastries each time. With the event *supply to aide*, the aide takes the tray and has all the single pastries at hand. He puts each single pastry onto one of the empty shelves.

Figure 9.10 models this behavior: The variable X varies freely over the multi-sets of pastries. A step of the event *bake* assigns such a multi-set, M, to X, causing the predicate *on counter* to apply to M. Notice, *on counter* does not apply to the single elements of M. As an example, suppose the predicate on counter applies to the multi-set {*bread, bread, pie*}. The transition *supply to aide(X)* then causes the predicate *aide with pastries* to apply to each single pastry, the two breads, as well as the pie. This is achieved by the *elm*-notation, as used in the edge inscription *elm(X)*: intuitively stated, the *elm*-notation turns a set into its single elements. In the dynamic trays example, this fits with the types of the predicates: the predicate *on counter* applies to sets of pastries; the predicate *aide with pastry* applies to single pastries.

Conceptually, the inscription *elm(X)* on the edge from the transition *supply to* aide(X) to the place *aide with pastry* is a shorthand for the predicate logical formula

$$\forall m \in M : aide \ with \ pastries(m). \tag{9.1}$$

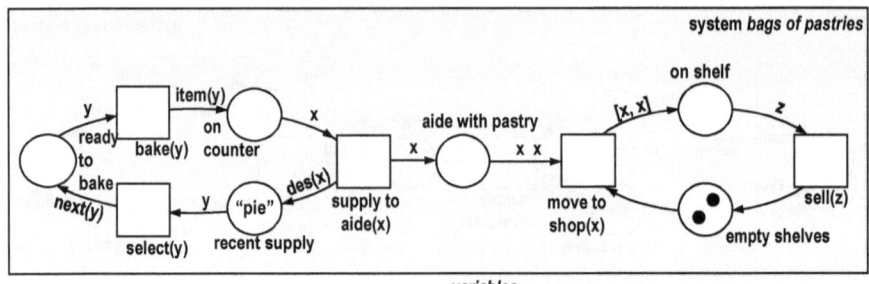

variables
x: pastries y: descriptions z: multiset of two pastries

structure *bags of pastries*		
domains pastries = {bread, cake, pie, rol, biscuit} descriptions = {"bread", "cake", "pie", "rol", "biscuit"}	*predicates* on counter, aide with pastry: pastries recent supply, ready to bake: descriptions on shelf: sets multisets of two equal pastries empty shelves: ℕ	*function* next: descriptions ⟶ descriptions next("bread") = "cake" next("cake") = "pie" next("pie") = "bread"
constants "pie": descriptions bread, cake, pie: pastries	*function* item: descriptions ⟶ pastries item("bread") = bread item("cake") = cake item("pie") = pie	*function* des: pastries ⟶ descriptions des(bread) = "bread" des(cake) = "cake" des(pie) = "pie"

Fig. 9.9 Bags of two alike pastries, and two alike shelves.

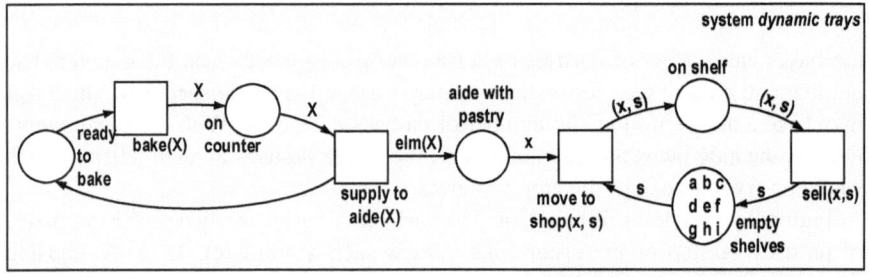

variables
s: shelves x: pastries X: sets of pastries

structure *dynamic trays*		
domains pastries = {bread, cake, pie} shelves = {a, b, c, d, e, f, g, h, i}	*constants* a, b, c, d, e, f, g, h, i: shelves *proposition* ready to bake	*predicates* on counter: multisets of pastries aide with pastry: pastries on shelves: set of pastries × set of shelves empty shelves: shelves

Fig. 9.10 Bakery system with trays.

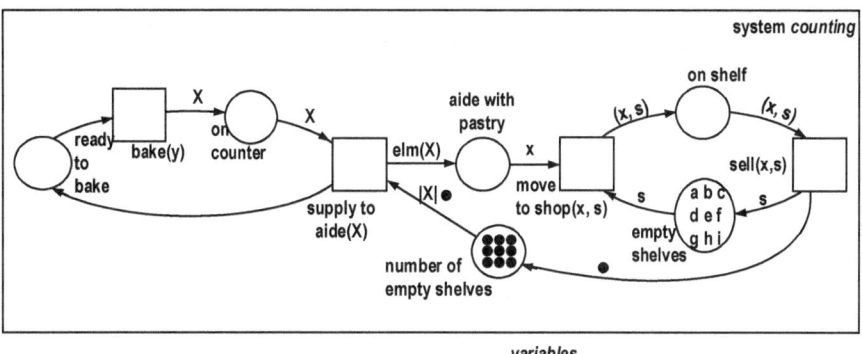

variables
s: shelves x: pastries X: sets of shelves

domains	**predicates**	**proposition**
pastries = {bread, cake, pie}	on counter: multiset of pastries	ready to bake
shelves = {a, b, c, d, e, f, g, h, i}	aide with pastry: pastries	
token = {•}	on shelf: set of (pastries × shelves)	**function**
	empty shelves: set of shelves	| ... | : multiset of pastries → ℕ
constants	number of empty shelves: multiset	| M | : number of elements of M
a, b, c, d, e, f, g, h, i: shelves	of tokens	

Fig. 9.11 Counting the empty shelves.

9.8 Trays with different number of pastries: counting

In the above *system dynamic trays*, the baker may bake any number of pastries. Unlimited numbers of trays may pile up on the *counter*, and unlimited numbers of pastries may pile up at *aide with pastry*. We want the transition *supply to aide(X)* to occur with an assignment $X = M$ only if enough shelves are empty to accept all pastries of M.

For the occurrence of the transition *supply to aide(X)*, a particular challenge is the unknown number of required empty shelves. Figure 9.11 solves this problem by two means: Firstly, *number of empty shelves* is a proposition that may hold many black dots; in fact, it receives a black dot whenever a pastry is sold. Secondly, the "number-of"-function "$|\cdot|$" assigns each set M the number $|M|$ of its elements. Hence, in Figure 9.11 with an assignment $X = M$, the term "$|X|$" denotes the number $|M|$ of black dots. Hence, with an assignment $X = M$, *supply to aide(X)* requires $|M|$ dots on *number of empty shelves*, and removes this number of dots from *number of empty shelves*.

This example shows the expressive power of structures in combination with states and events to describe dynamic behavior.

Chapter 10
Representing it all symbolically

The previous Part III of this monograph describes in detail the concept of *structures*, consisting of *domains*, *constants*, *functions*, *predicates*, and *propositions*. Additionally, *signatures* and their *terms* are used to represent structures and their components. This framework is now applied to general system modules, leading to the notion of *system schema*. Such a schema does not represent just one system, but many similar ones, as defined in Section 7.2, pages 74f.

10.1 The bakery in symbols

In the previous Chapter 9, each general system module comes with its own structure. In many cases, however, we are interested in all systems with similar structures, as defined in Definition 7.2, page 74, in Section 7.4. For example, the *three pastries* structure of Figure 6.1, page 70, and the *five pastries* structure of Figure 6.2, page 71, are similar: They are both instantiations of the *bakery* signature in Figure 7.1, page 74. Many of the structures described in the previous Chapter 9 are instantiations of the signature *bakery*. Figure 10.1 shows the symbolic representation of the basic bakery business. To emphasize the schematic aspect, all denotations, inscriptions, et cetera are underlined.

10.2 System schemata and their interpretation

The above Definition 8.1, page 91, of general system modules in Section 8.3 assumes a signature Σ and a Σ-structure S, taking predicates of S as places. The terms in edge expressions and the constant symbols in places are intended to be interpreted over S. We now refrain from focusing on *one* interpretation S, and consider *any* interpretation of Σ: assuming just a signature Σ, we take the *predicate symbols* of Σ as places.

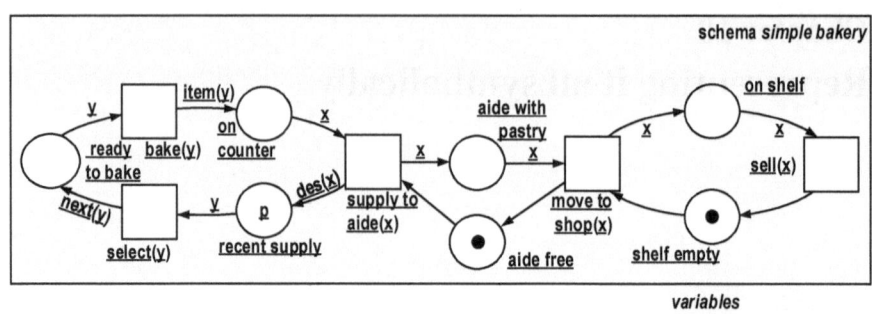

variables
x: pastries y: descritions

signature *simple bakery*

domain symbols
pastries
descriptions

predicate symbols
on counter, aide with pastry, on shelf: pastries
recent supply, ready to bake: descriptions

function symbols
des: pastries ⟶ descriptions

constant symbol
p: descriptions

proposition symbols
aide free, shelf empty

next: descriptions ⟶ descriptions

item: descriptions ⟶ pastries

Fig. 10.1 The symbolic representation of the bakery.

Definition 10.1 (system schema) Let Σ be a signature and let X be a set of Σ-variables. Let N be a net module where:

- each place of N is a predicate symbol of Σ, inscribed with constant symbols or a term of the form $elm(a)$, where a is a constant symbol for sets.
- each edge (p, t) or (t, p) of N is inscribed with a term u over Σ and X, where the type of u coincides with the type of p. Those terms may also include $elm(a)$, where a is a constant symbol or a variable for sets.

Then N is a *system schema over Σ and variables X*.

The above Figure 10.1 shows an example.

Definition 10.2 (interpretation of a system schema) Let Σ be a signature, let X be a set of variables, let N be a system schema over Σ and X, and let S be a Σ-structure. Let N_S be the general systems module where:

- the underlying net modules of N and N_S are identical;
- each place p of N_S is the interpretation of the predicate symbol p of N in S;
- each place inscription of N_S is the interpretation of the place inscription of p of N in S.

Then N_S is the *S-interpretation* of N.

For example, the *bakery with pastries* system of Figure 8.12, page 93, is an S-interpretation of schema *bakery* of Figure 10.1 with S the structure *three pastries* of Figure 6.1, page 70.

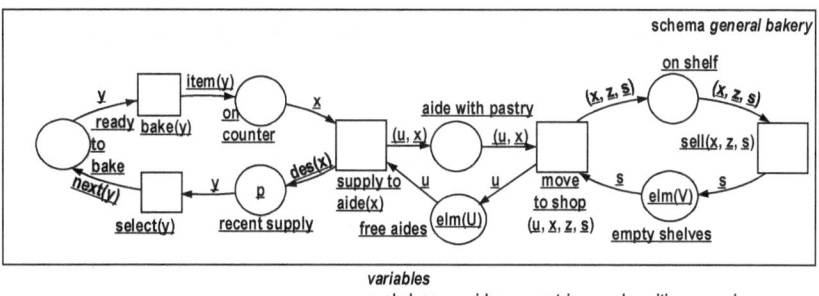

Fig. 10.2 A schema for the bakery.

We now consider the variant where the number of aides and of shelves remains open and is fixed by each instantiation anew. In Figure 10.2 the symbol U stands for a set of aides; for example, the set {*Alice, Bob*} of Figure 9.6, page 101. The *elm*-notation guarantees that the *free aides* predicate applies initially to each single element of this set, but not to the set itself! Likewise, the symbol S stands for a set of shelves.

HERAKLIT and Petri nets

HERAKLIT employs Petri nets for the description of behavior. Nevertheless, HERAKLIT deviates from usual Petri net dynamics in several aspects. The literature on Petri nets usually defines global states, called *markings*, where a marking A assigns a set of "black" tokens or (composed) items to each place p of a Petri net N. Occurrence of a transition t then diminishes or increases the number of tokens on places in the vicinity of t, thus leading to a fresh marking, A'. Furthermore, an initial marking is assumed. Altogether, this amounts to a setting in the style of automata theory. In contrast, HERAKLIT avoids any global aspect. A HERAKLIT run of a Petri net comes without global states. Consequently, each run R of a module M is also a run of $M \bullet N$ and of $N \bullet M$.

Related work to Part IV

In the short history of informatics, computability theory has frequently been conceived as *the* theoretical basis. Nevertheless, numerous proposals challenge this assumption. A classical reference is [94]. This paper as well as [4, 53], but also many others, proposes to transcend computability theory by emphasizing *interaction* as a fundamental computing feature, and asking for a most abstract model of interacting computers. Despite presenting a more liberal view, most of these approaches conceive computer embedded discrete systems from the perspective of computers and software. Meanings of systems is defined in a formal setting, not in terms of the real world. Here we glance at a few of them:

[15] suggests a formal framework for systems and their composition. Interfaces consist of channels for streams of messages or actions. Each interface handles either input streams or output streams. A composition operator is defined for the case of two systems with disjoint input and disjoint output channels. This operator is later specialized in several ways.

The *actor model*, introduced by Gul Agha in [1], treats *actors* as basic building blocks of concurrent computation. An actor responds to messages it receives by making local decisions, creating more actors, sending more messages, and modifying its private state. The dynamic creation of new actors in the actor model prevents modeling actors immediately as HERAKLIT modules.

Farhad Arbab's channel-based coordination model Reo [3] applies structured *connectors* to compose components. The simple versions of these connectors can immediately be modeled by HERAKLIT gates. More involved connectors, as well as composition patterns, can be described as HERAKLIT adapters.

Robin Milner suggested *bigraphs* [59] as a variant of process algebras, respecting and representing aspects of locality and connectivity. A bigraph integrates various aspects at once, whereas a HERAKLIT model considers architecture, data, and control separately, before integrating them.

Many fewer approaches explicitly take the perspective of the real- or imagined-world as the starting point for static and behavioral descriptions of computer embedded systems.

Gurevich's *abstract state machines* (ASM) [34] start out with Σ-structures as discussed in Chapter 7.2, pages 74ff. An ASM program M updates a given actual Σ-structure by evaluating the meaning of some Σ-terms. A step of M consists of a structure and its updated version. A *single run* is a sequence of updated Σ-structures. Parallel and distributed ASM versions generate runs which update sets of Σ-terms. This concept, indeed, allows for a formal notion of algorithm in the real or imagined worlds.

ASM represent architecture by conventional software structures. *Gurevich* characterizes the expressive power of ASM by means of three "postulates". *Boker* and *Dershowitz* in [13] add a fourth postulate, and formulate a *Church-Turing*-Thesis analogy for Σ-structures with "effective" basic functions. Here, the question occurs of how these results can be exploited to characterize the expressive power of HERAKLIT behavioral modules. And vice versa, it would be interesting to identify particular HERAKLIT behavioral modules that are equivalent to *abstract state machines* (ASM). Gurevich's postulates have been reformulated for Petri net schemata in [75]. It remains an open problem to include arguments on architecture in these postulates.

Bjørner's domain modeling [12] suggests an ontology of *domains*, where a domain is essentially a (huge) Σ-structure, for an alphabet Σ. Similarly to ASM, *domain modeling* formulates behavior by updates of Σ-terms, and single runs as sequences of Σ-structures. Parallel and distributed versions generate runs which update sets of Σ-terms. Domain modeling represents architecture by explicit CSP-like channels. Elements of a domain have similar properties; so, they can then be analyzed in a systematic manner, and their behavior can be defined systematically. This supports deep insight into the path from informal to formal arguments. It would certainly be very useful to refine HERAKLIT by *Bjørner's* sophisticated and refined domains.

In the area of Petri nets, *coloured Petri nets* (CPN) [43] may be the closest to HERAKLIT. However, CPN comes with classical, finite data structures. Each net operates with one structure; there is no schematic level, and no specific composition mechanism for modules.

As a typical potential application scenario for HERAKLIT, we consider the paper [78]. This paper provides concepts to define the integration of enterprise applications on a formal basis. To this end, a number of formal frameworks are employed. It may be useful to take HERAKLIT-based concepts as such a formal basis.

From the perspective of business informatics and business process management, the need for integrated modeling of enterprises has often been emphasized, e.g. [22, 29, 64, 83]. In addition, the need for conceptual modeling for practical purposes and objectives is well recognized, e.g. [30, 74, 81, 88, 93]. All these ideas and concepts seem to offer promising venues for understanding computer-integrated systems. However, these and other approaches typically lack a strong theoretical foundation. In other words, their bold and far-reaching speculations about possible domain-oriented approaches to computer-integrated systems miss precise and explicit foundations.

HERAKLIT provides this kind of foundations. We refrain from enumerating further potential application scenarios for HERAKLIT, as this would exceed the related work section of this book.

Part V
HERAKLIT in action

Chapter 11
Case study: retailer

A comprehensive case study exemplifies the construction of *large* HERAKLIT models. We model the central business processes of a retailer and its business network partners, we show how they can be composed and abstracted, and we describe their behavior. From the constructs of this case study the reader can intuitively understand the interplay of HERAKLIT concepts.

The case study is divided into eight sections. Section 11.1 introduces the content of the case study and the developed model. A HERAKLIT model is basically divided into individual modules which are presented in Section 11.2. Section 11.3 explains how data and objects are modeled in the case study. The concept of runs is exemplified in Section 11.4. The introduced concepts are transferred to schemata to talk about arbitrary data, objects, and runs in Section 11.5. Section 11.6 then details the individual modules in terms of their structure and behavior. The complete model overview is shown in Section 11.7, while Section 11.8 explains how further views on the complete model can be described.

11.1 Content of the case study and the developed model

The retailer of the case study

A retailer sells articles to its customers via an online shop. No articles are sold via other forms of distribution such as stationary retail, wholesale, or agents.

The retailer has three types of business network partners:

- its *customers*: they send in orders; the retailer informs the customers about delivery dates, et cetera;
- the *supplier*: the retailer reorders out-of-stock goods; the supplier delivers ordered goods to the retailer;
- the *freight forwarders*: they receive packed freight from the retailer and deliver it to the customers.

P. Fettke, W. Reisig, *Understanding the Digital World*,
https://doi.org/10.1007/978-3-031-61898-7_11

The retailer itself is divided into three parts:

- The *order management* receives orders from customers, asks the inventory management to confirm the availability of each ordered item and issues delivery orders to the warehouse. If necessary, the order management system initiates partial deliveries.
- *Inventory management* keeps a list of the inventory in the warehouse. If an item's available quantity becomes too low, inventory management reorders from the supplier and asks the warehouse to acknowledge the receipt of these items.
- The *warehouse* packs ordered goods according to the delivery orders of the order management and hands over the freight to a freight forwarder. It also receives goods from the supplier, sorts them into the warehouse and notifies the inventory management system that the goods have been received.

The retailer and its partners form a system that must function correctly as a whole. For example, all ordered items should be delivered to a customer, if necessary in partial deliveries. Conversely, only those articles that the customer has ordered should be delivered. Within the retailer, inventory management should only prompt the warehouse to deliver if the necessary articles are available in the warehouse.

The HERAKLIT model for the retailer and its business network partners

We construct a HERAKLIT model for the entire system, containing the described retailer and its three business network partners. First we model the abstract structure of the whole system and the three business network partners of the retailer as *HERAKLIT modules* in the context of their principal interaction. This is followed by a detailed and systematic presentation of the data and objects in an abstract version. The representation on a schematic level is left to concrete *instantiations* which specify how exactly customers, articles, goods, freight, et cetera look. Finally, the behavior of the individual modules is represented in the form of local steps and mathematical operations on the data and objects involved.

The model leaves a number of decisions open. In particular, order management

- can bundle available items in the warehouse into partial deliveries according to particular choices;
- can select different goods in the warehouse for the same ordered item;
- does not determine which of the available freight forwarders will deliver.

Thus, the model precisely captures the aspects of typical trading operations.

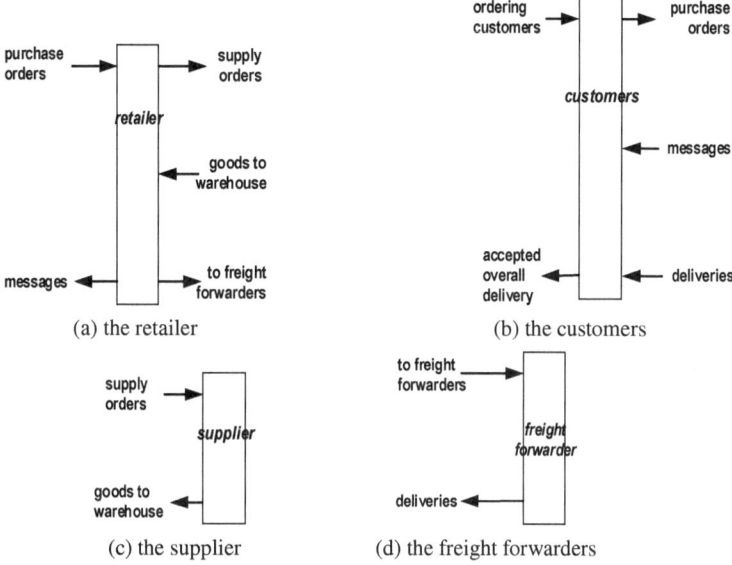

Fig. 11.1 Abstract HERAKLIT modules: the retailer and its three kinds of partners.

11.2 Architecture

The modules

The module in Figure 11.1a shows the architecture of the retailer. The interior of the *retailer* is shown inside the module; here we abstract completely from internal details and only write the name of the module. The interface consists of five gates with the labels *purchase orders* (from customers), *messages* (to customers), *supplier orders* (to the supplier), *goods to warehouse* (from the supplier), and *to freight forwarders* (to freight forwarders). The gates are represented as arrows, indicating the flow of goods or data.

Similarly, the modules of the Figures 11.1b, 11.1c, and 11.1d show the three types of business network partner: *customer*, *supplier*, and *freight forwarder* with their respective gates. Figure 11.2 shows the three departments of the retailer, again as HERAKLIT modules.

All HERAKLIT modules in Figures 11.1 and 11.2 are *abstract* modules: Their interior contains only the name of the module. Typically, however, the interior of a HERAKLIT module shows details of its structure or behavior. The structure of a module is often a composition of sub-modules.

Each gate of the modules in Figures 11.1 and 11.2 is – intuitively formulated – an *entrance* or an *exit* through which objects, documents, information, et cetera flow. The direction of this flow helps us to understand the behavior of the module; we indicate it with an arrowhead. Such arrowheads are part of the labels of the gates.

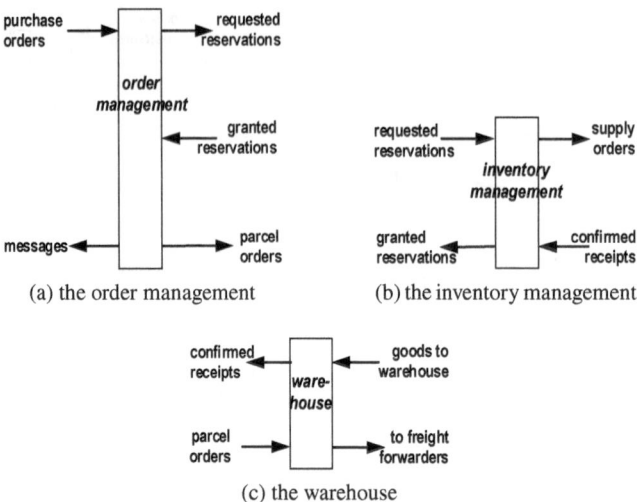

(a) the order management   (b) the inventory management

(c) the warehouse

Fig. 11.2 Abstract versions of the three departments of the retailer.

The composition of modules

Figure 11.3a composes the *order management* and the *inventory management* from Figures 11.2a and 11.2b. Both have gates with the label *requested reservations*. In Figure 11.3a the two gates merge into one element inside *order management • inventory management*. The same applies to the two gates with the label *granted reservations*. The other three gates of *order management* and the other two gates of *inventory management* then form the interface of the composed system *order management • inventory management*.

Figure 11.3b depicts the system

$$inventory\ management \bullet warehouse .\qquad(11.1)$$

Both composed modules can now be extended by the missing third department of the retailer. As a result, the following two modules can be composed:

$$(order\ management \bullet inventory\ management) \bullet warehouse\qquad(11.2)$$

and

$$order\ management \bullet (inventory\ management \bullet warehouse).\qquad(11.3)$$

These two modules are identical; therefore, in Figure 11.3c the brackets can be omitted.

Figure 11.4a composes the retailer module with the modules for the customers, the suppliers, and the freight forwarders to form the *overall module* of the case study. The modules involved are then *sub-modules* of the overall module.

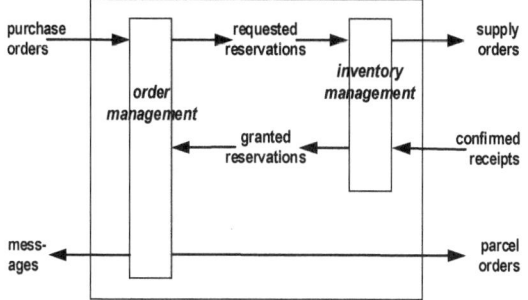

(a) the composition *order management • inventory management*

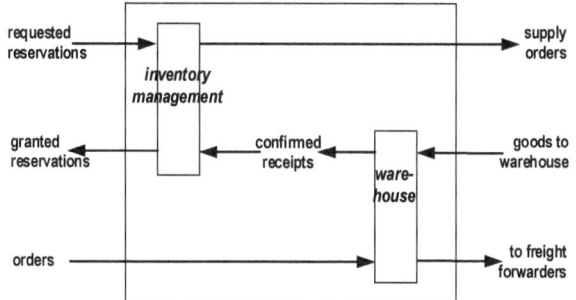

(b) the module *inventory management • warehouse*

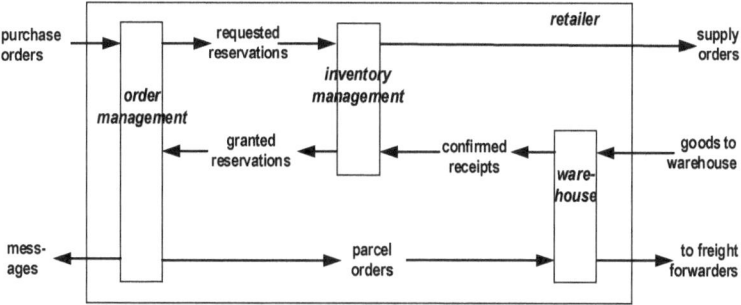

(c) the retailer, defined as *order management • inventory management • warehouse*

Fig. 11.3 Composition of departments.

Each gate of the retailer module is by design also a gate of one of its three sub-modules. Therefore, in Figure 11.4a the retailer module can be replaced by its three sub-modules, as in Figure 11.4b. An arrowhead at a gate intuitively indicates a flow direction. Such arrowheads are also intuitively useful in composed modules.

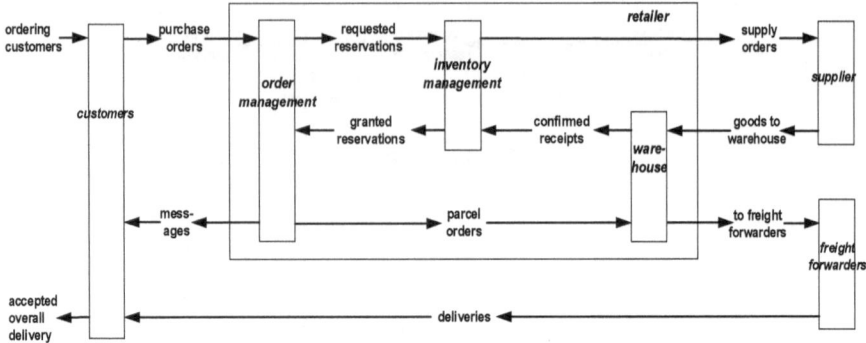

(a) the composition *customers • retailer • supplier • freight forwarders*

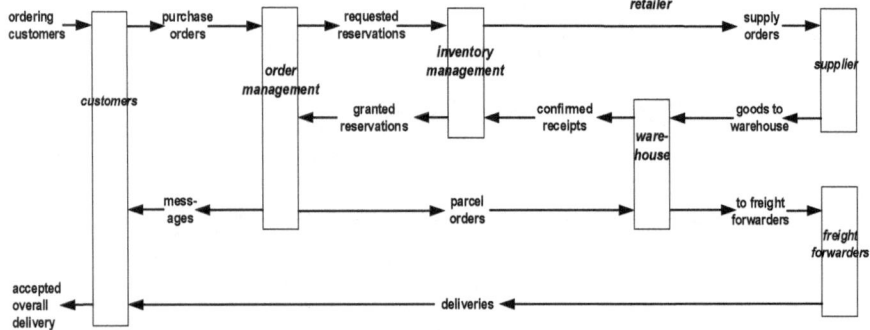

(b) the composition *customers • order management • inventory management • warehouse • supplier • freight forwarders*

Fig. 11.4 The retailer in the context of its business network partners.

Abstraction of modules

The module in Figure 11.3c is not abstract, but has an inner structure. If one abstracts from this inner structure, one obtains Figure 11.1a, the *abstract version* of the retailer. In general, each module has its uniquely determined *abstract version*: The surface remains the same, but the interior is deleted except for the name of the module. For a module L, $[L]$ denotes its abstract version. Figure 11.3c is named *retailer*; therefore – strictly speaking – Figure 11.1a must be named *[retailer]*. But now one can see that the modules in Figure 11.1 are all abstract; therefore, the representation remains unique even without the abstraction operator $[\cdot]$. If one forms a module as a composition of given modules and would like to refer to the abstract version, the abstraction operator removes ambiguity. An example is shown in the comparison of Figure 11.4a with Figure 11.5.

The transition from a module L to its abstract version $[L]$ can be reversed and L can be seen as one of many possible refinements of $[L]$. For example, Figure 11.3c

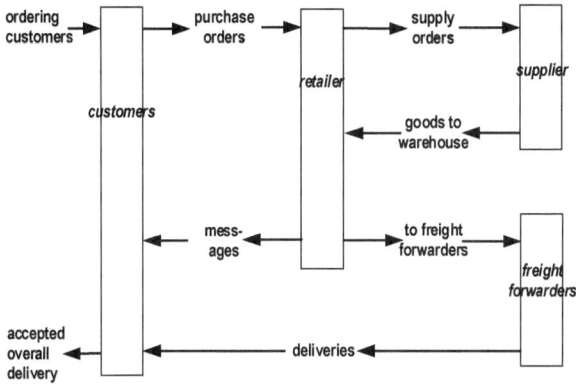

Fig. 11.5 Composition of abstractions: [*customers*]•[*retailer*]•[*supplier*]•[*freight forwarders*].

is a refinement of Figure 11.1a. If in a composed module $L = L_1 • L_2 • ... • L_n$ a sub-module L_i is refined or abstracted, nothing changes in the other sub-modules.

The three sub-modules depicted in Figure 11.3c will be further refined later, as will the business network partners given in Figure 11.1.

Views

Abstraction and composition of sub-modules result in different views of the overall system. In particular, each business network partner has its own view, namely the abstraction of the other composed sub-modules. Figure 11.6 depicts these three views.

A module (at any level of detail) *sees* the rest of the system generally in two parts: the sub-module on its left and the one on its right. Figure 11.7a shows this perspective of the retailer. But each of its sub-modules also has its own perspective, as shown in Figures 11.7b, 11.8a and 11.8b.

11.3 Statics

Data, things, types, multi-sets

We examine some data aspects of the case study in more detail. An *order* consists of a *customer* (more precisely: a customer name) and an *article list*. This in turn is a set of *items*. And each item consists of an *article* and the *number* of ordered units.

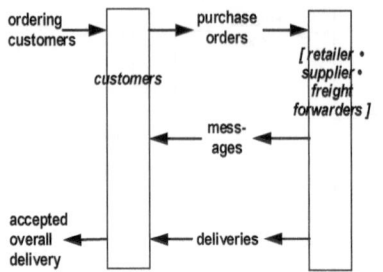

(a) the system from the customer's perspective:
customers •[retailer• supplier•freight forwarders]

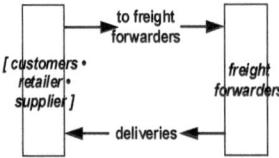

(b) the system from the perspective of the freight forwarders:
[customers • retailer• supplier] •freight forwarders

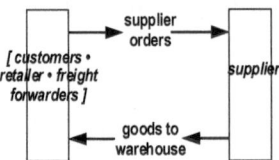

(c) the system from the supplier's perspective:
[customers • retailer•freight forwarders] • supplier

Fig. 11.6 The system from the perspective of the three business network partners.

Customer names, articles, article positions, and article lists are *data*; one can display them in a catalog, send them digitally, print them on paper, et cetera. In particular, they can be processed in digital form by a computer.

In addition, there are concrete real-world *things*, in the example the *shoes, pants, shirts*, and *hats*, that are the goods that the supplier sends to the warehouse, which the warehouse packs and hands over to the freight forwarders as freight, which is then delivered to the customers.

Data and things have very different properties: A data element such as an article list can be easily copied or disassembled; an object cannot. One can do little more with an object than bundle it together with others in freight, transport it from place to place, and then open the bundle again. From this follows an interesting property that data does not have: An existing thing, for example a hat, is always in exactly one place. Data and things are *objects*, which HERAKLIT represents in the same formalism.

In the systematic structure of HERAKLIT, objects are elements of *sets*. They are, therefore, not to be understood in the special sense of object-oriented modeling or

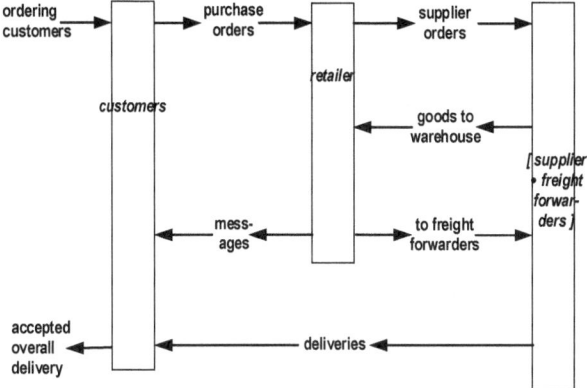

(a) perspective of the retailer: [*customers*] • *retailer* • [*supplier* • *freight forwarders*]

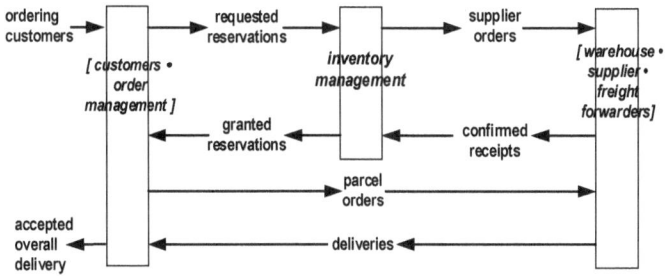

(b) perspective of inventory management: [*customers* • *order management*] •
inventory management • [*warehouse* • *freight forwarders*]

Fig. 11.7 Perspectives of the retailer and its departments (1/2).

programming. Often objects are *multi-sets*. In a multi-set, an element can occur
more than once: for example, a delivery for an order can contain two hats and three
pairs of shoes without distinguishing between the two hats or the pairs of shoes.
We write this order as multi-set [*hat, hat, shoes, shoes, shoes*]. Formally, a multi-set
with elements from a set A is a mapping $M : A \to \mathbb{N}$ which assigns each element of
A its number of occurrences in M; in the example, $M(hat) = 2$ and $M(shoes) = 3$.

Multi-sets L and M of a set A can be added:

$$(L + M)(a) =_{\text{def}} L(a) + M(a), \tag{11.4}$$

multiplied by a scalar (a natural number n)

$$(n \cdot M)(a) =_{\text{def}} n \cdot (M(a)), \tag{11.5}$$

and compared with each other:

$$L \leq M =_{\text{def}} \text{ for all } a \in A : L(a) \leq M(a). \tag{11.6}$$

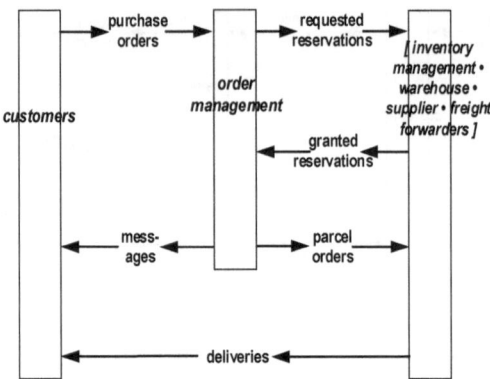

(a) perspective of the order management: [*customers*] • *order management* • [*inventory management* • *warehouse* • *supplier* • *freight forwarders*]

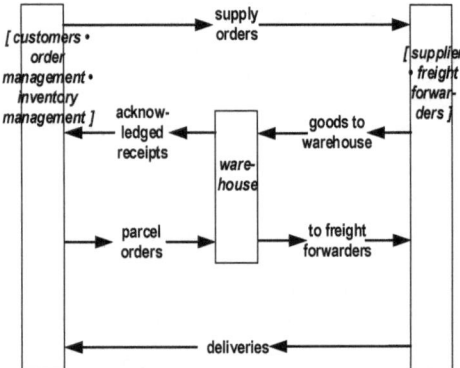

(b) perspective of the warehouse: [*customers* • *order management* • *inventory management* • *warehouse* • [*supplier* • *freight forwarders*]

Fig. 11.8 Perspectives of the retailer and its departments (2/2).

The power set $P(M)$ is the set of all sets N with $N \leq M$. $M(A)$ is the set of all multi-sets over A.

Finally, we use *predicates*: A predicate p either applies to an element, or it does not. For a set M we write $p(elm(M))$ if p applies to every element of M.

Figure 11.9 summarizes the notations used in this case study.

The structure S and the schema of the retailer

The model of the retailer consists of five different types of objects:

- *basic domains*: these are *customers, articles, dates, goods, freight forwarders*;

ground sorts
N the set of natural numbers

derived sorts
A × B *Cartesian product of sets **A** and **B***
P(A) *power set of set **A***

function symbols
_+_ : addition on N
_-_ : subtraction on N
_<_ : order on N

multi-sets
intuitively: a *multi-set* is a collection of elements,
where an element may occur more than once.
Notation: for example, [a, a, b] or {a, a, b}
Formally: for a given finite set A, a *multi-set M over A*
is a mapping M: A → N.

operations on multi-sets
for multi-sets L and M:
- Sum (union) L+M, with (L+M)(a) = L(a)+M(a)
- For n ∈ N: n•M: n-fold addition of M,
- partial order L < M: component-wise,
P(M) power set of of M (set of all multi-sets N ≤ M)
M(A) set of all multi-sets over A

a notation for predicates, p
for a set A: "*p applies to elm(A)*" is shorthand for "*p
applies to each element of A*".
For multi-sets: p applies to all instances.

Fig. 11.9 HERAKLIT multi-sets.

- *derived domains*: an *article position* consists of an article and a quantity, an *article list* is a multi-set of article positions, an *article set* is a multi-set of articles, and a *goods set* is a multi-set of goods;
- *constants* that explicitly appear in the model: specific items, customers, and initially listed items, goods, and freight forwarders;
- *functions*, which relate basic and derived domains to each other: each article position *p* together with an article list *a* corresponds to a multi-set *p* of articles *a*, and each good *w* corresponds to an article $f(w)$;
- *predicates* are derived from the basics: *ordering customers, sent orders, copies of sent orders, delivered goods*, et cetera are predicates. A predicate either applies or does not apply to an element of a set. Example: The predicate *sent orders* applies to all orders that have been sent by customers but have not yet been processed by the retailer.

We also use a number of variables for the different domains. Basic and derived domains, as well as functions and predicates over these domains, form a *(Tarski) structure*. Such structures form the basis for the formulation of dynamic behavior with HERAKLIT behavior models.

Figure 11.10 depicts the structure we will use to describe the case study below.

11.4 Dynamics: networked runs

Customers submit purchase orders to the retailer and receive deliveries from freight forwarders; the retailer reorders any out-of-stock items from the supplier, it assigns a freight forwarder to deliver freight from the supplier to the customer, et cetera. Such

ground sorts
KN = {Ute, Max}   *customers*
AR = {"shoes", "hat", "pants"} *articles*
WA = {shoes, hat, pants}   *goods*
TE   = {12/23, 12/24}   *dates*
SP   = {Maier, Müller, Schulz} *freight forwarders*

derived sorts
AP = AR × N   *items*
AL = *M*(AP)   *article lists*
AM = *M*(AR)   *sets of articles*
WM = *M*(WA)   *sets of goods*

functions
f(w) = "w", for w ∈ **WA**
f([a1, ... , an]) = [f(a1), ... , f(an)] for [p1, ... , pn] ∈ **AM**

(a,n)' = n[a] for (a,n) ∈ **AP**
[p1, ... , pn]' = p1' + ... + pn' for [p1, ... , pn] ∈ **AL**

examples
p1' = {"shoes", "shoes"}
[p1, p2, p2]' = {"shoes", "shoes", "hat", "hat"}

constants
p1: AP = ("shoes", 2),
p2: AP = ("hat", 1)
K: P(KN) = {Ute, Max}   *ordering customers*
G: AL = {("shoes", 3), ("hat", 1)}   *initially listed articles*
H: WM = {shoes, shoes, shoes, hat}   *initially listed goods*
R: P(SP) = {Maier, Müller} *initally available freight forwarders*

Fig. 11.10 The HERAKLIT structure *S* of the retailer.

behavior, composed of individual events that are considered elementary, is described in HERAKLIT *behavioral modules*.

Local states and events

In the natural sciences and engineering, the behavior of a system is very often modeled as a continuous process along a time axis of real numbers. Here we understand *behavior* fundamentally differently: The behavior of a system is described by *local states*, which are updated by discrete *events*. The result of an event can be the cause of further events.

The description of local states is based on predicates (see Figure 11.23, page 141): A *local state* is a predicate *p* together with an object *o*. Dynamics are created when the object *o reaches* the local state; then *p* applies to *o*. If *o leaves* the local state, *p* no longer applies to *o*. Reaching and leaving local states is synchronized by the occurrence of *events*: Some objects reach or leave some local states.

A typical example is the ordering of items: For example, when the customer *Ute* submits an order, she leaves the local state *ordering customers* and the order reaches the two local states *submitted orders* and *copies of submitted orders*. Figure 11.11 graphically represents the event *Ute submits purchase order {p1, p2, p3}*: Each of the three ellipses represents one of the involved predicates together with the corresponding object to which the predicate applies. The rectangle *a* contains the name of the event. An arrow between an ellipse and the rectangle indicates whether the object leaves (arrowhead on the rectangle) or reaches (arrowhead on the ellipse) the corresponding local state when the event occurs.

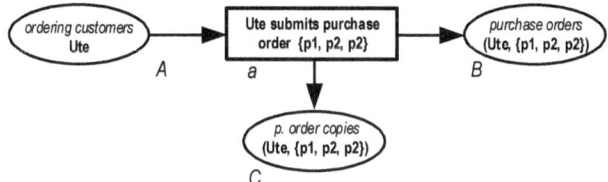

Fig. 11.11 Event *Ute sends order {p1, p2, p3}*. Initially, the predicate *ordering customers* applies to the customer *Ute*, and the predicates *purchase orders* and *purchase order copies* apply to no purchase orders. After occurrence of *Ute submits purchase order {*p1, p2, p2*}*, Ute is no longer a *submitting customer*, and both the predicates *purchase orders* and *purchase order copies* apply to the *purchase order* (Ute, p1, p2, p2).

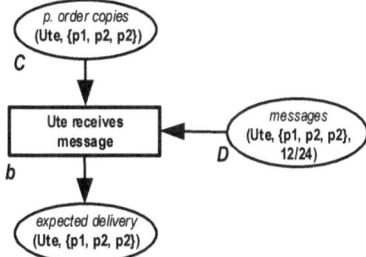

Fig. 11.12 Event *Ute receives message*: delivery arrives on December 24th. With a copy of the submitted purchase order and a corresponding message (including a delivery date), the event *Ute receives message* occurs, resulting in the expected delivery *(Ute, {p1, p2, p2})*.

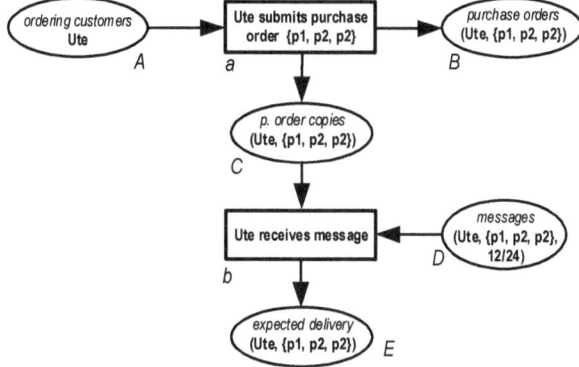

Fig. 11.13 Composition: merging of local states. Composition *e • f* of the two events *e = Ute submits purchase order {p1, p2, p2}* and *f = Ute receives message* by merging the two local states labeled *partial order copies (Ute, {p1, p2, p2})*.

Formally the event *Ute submits purchase order {p1, p2, p2}* has the structure of a *net*: each ellipse is a *place*, the rectangle is a *transition*, and the arrows form the *flow relation*.

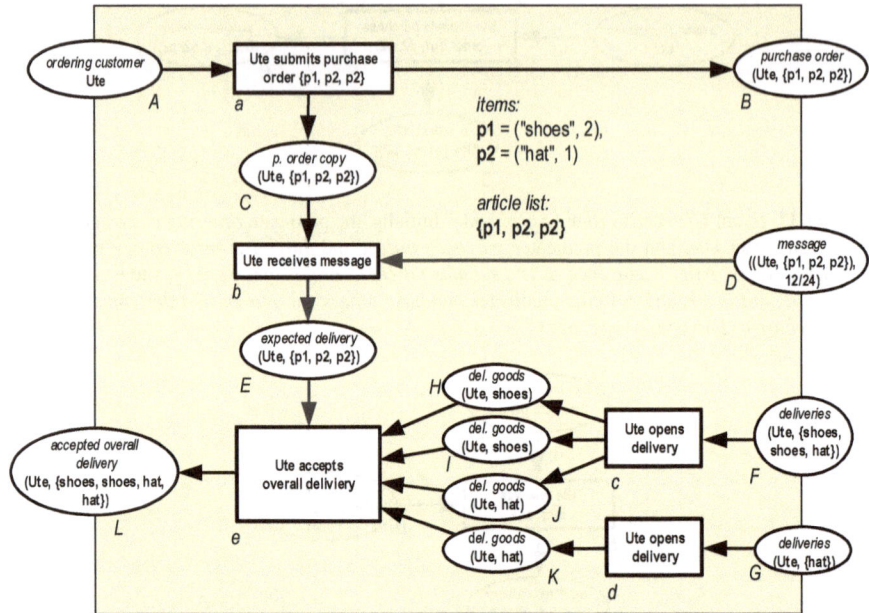

Fig. 11.14 A run of the module *customers*. After a purchase order is submitted, a message and two deliveries will eventually reach the module along its right interface. Ute will open both partial deliveries. Altogether, they contain two pairs of shoes and two hats. Ute accepts the goods as the expected overall delivery.

Two events *e* and *f* can be *composed* to *e • f*, if an object *o* reaches a local state *p* by occurrence of *e* and then leaves it with occurrence of *f*. As an example, Figure 11.12 shows the event *Ute receives message*. As shown above, the order (*Ute*, {*p*1, *p*2, *p*2}) reaches the local state *purchase order copies* (abbreviated as *p. order copies*) with the event *Ute submits purchase order* {*p*1, *p*2, *p*2} and leaves with the event *Ute receives message*. The two events are composed by merging the two instances of the local state *purchase order copies* as shown in Figure 11.13.

As with a single event, the composition of two events has the structure of a net, consisting of places, transitions, and arrows.

Runs of modules

The composition of several events describes *complex behavior*; a *run* is created. The composition in Figure 11.13 describes an initial part of a possible *run* of the customer module. Figure 11.14 extends this initial part to a complete run of the customer module. A module can behave in different ways; for example, the customer Ute can choose between very different article lists for her order.

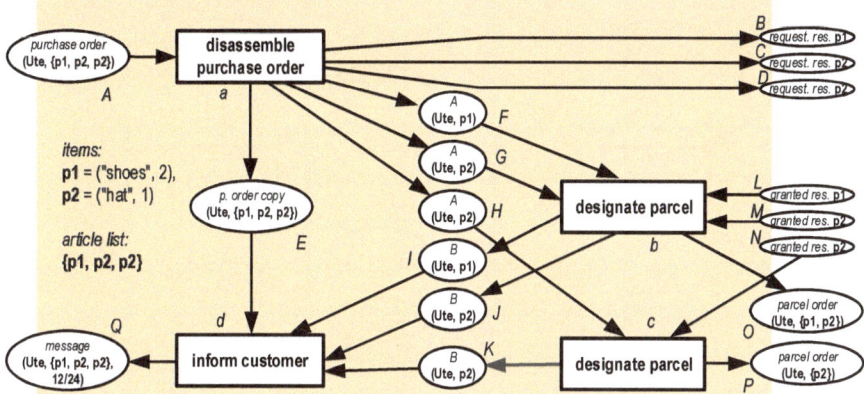

Fig. 11.15 A run of the module *order management*. (1) The order management disassembles the article list of the purchase order into its three items, sends for each item a request for reservation to the inventory management (without naming the client), and retains a copy of the purchase order. (2) Upon receiving grants for the requested reservations, the order management designates two parcels and sends corresponding orders to the warehouse. (3) Finally, the customer *Ute* is informed about the date of the last partial delivery, 12/24.

For the run of the module *order management* in Figure 11.15 no sequential order is assumed in which the confirmations of the reservation of the article positions reach the module or in which the module orders the two parcels. However, the customer will be notified only after it has been ensured that the retailer has the appropriate goods in stock for each item position ordered.

The *inventory management* module must answer a request for two pairs of shoes and a request for two hats. In the run shown in Figure 11.16, we assume that the list of available goods of the retailer indicates three pairs of shoes and one hat in stock (this list is for short, and without technical precision, called a *database*). Thus, a hat must be obtained from the supplier.

Figure 11.17 shows that the warehouse receives two hats from the supplier and reports this delivery to inventory management. In the run shown, two orders are completed: The first order includes one of the two delivered hats as a parcel and transfers the parcel as freight to the freight forwarders. For the second order, two pairs of shoes and a hat are in stock; they are packed and transferred to the freight forwarders.

As shown in Figure 11.18, the supplier receives the order for two hats from inventory management and delivers the hats to the warehouse. As Figure 11.19 shows, the two parcels are delivered independently to the customer *Ute*.

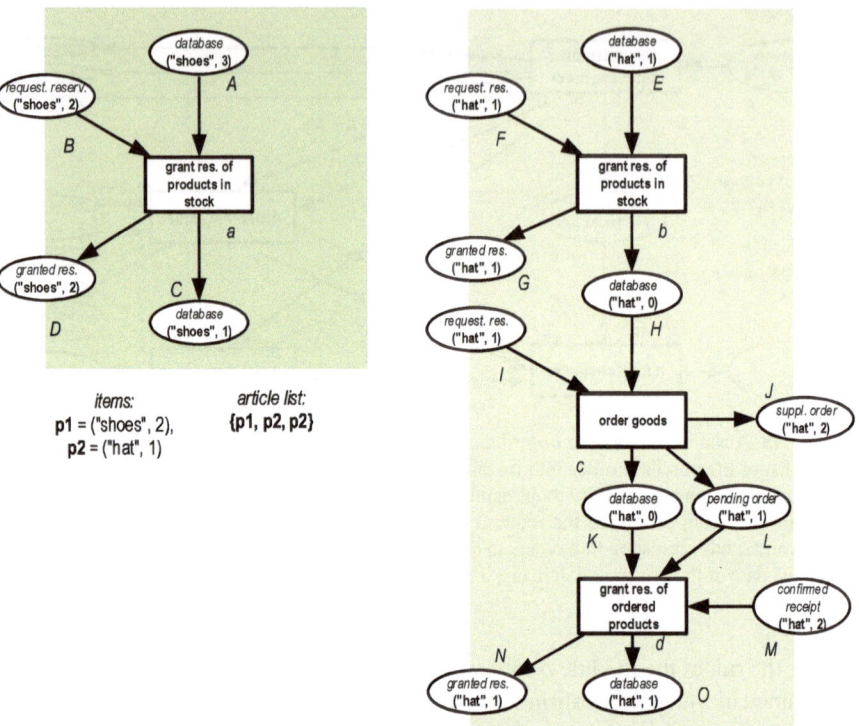

Fig. 11.16 A run, consisting of two parts, of the module *inventory management*. The inventory management acknowledges the reservation of two pairs of shoes and one hat, and updates the corresponding entries in the database. For the second required hat, two hats are ordered from the supplier – one hat for stock. After the warehouse's acknowledgment of receiving the two ordered hats from the supplier, inventory management informs order.

Composition of the runs of the modules: a run of the retailer

In the previous sections, we have seen how individual events can be composed to represent a possible behavior of each of the six modules. Now we compose the behaviors of the modules into a run of the retailer. This run starts with a customer's order, continues through all six modules, and finally returns to the customer with the delivery of the ordered goods. Figure 11.20 shows the composition of the processes of the six modules from the previous section. Two problems arise: First, the graphical arrangement of the interface elements to be merged generally does not match. Second, some interface elements whose labels do not match should be merged. Technically, we organize this with *adapter modules*, graphically represented as a line with a black square, called modules *p1-shoes2* and *p2-hat1* respectively.

As a graph, Figure 11.20 is *acyclic*: No arrow chain closes to a circle. Some events are thus ordered in a *before-after*-relationship: If an event e precedes an event f by a chain of other events, then f is certainly not before e. However, two events

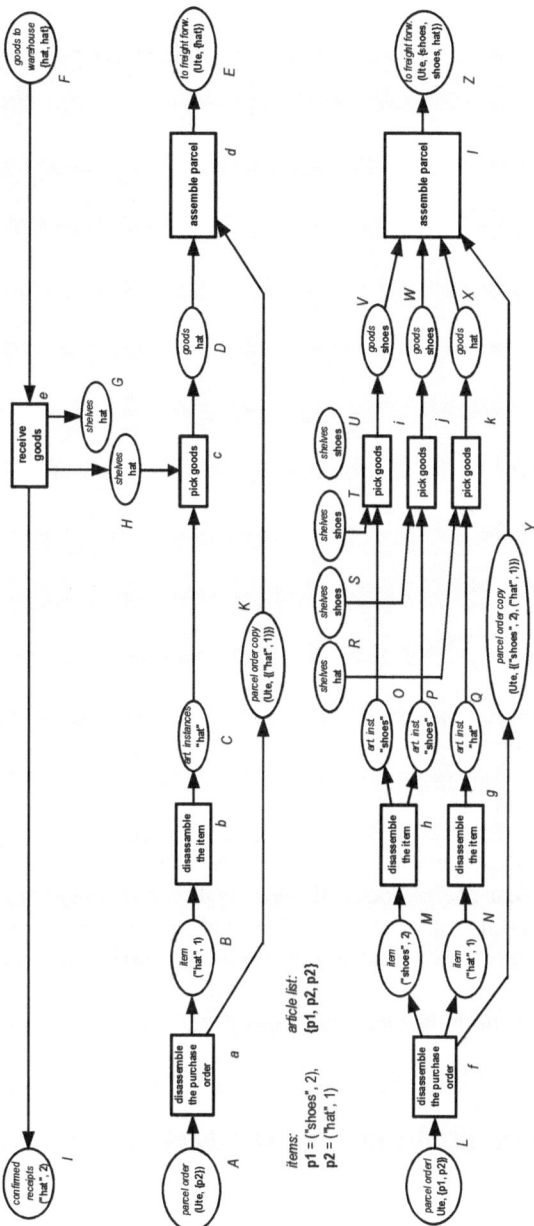

Fig. 11.17 A run of the module *warehouse*. The warehouse serves two orders with two detached sub-runs. Each of the two sub-runs disassembles an order into its items, and then generates a local state for each ordered article instance. For the lower order, the warehouse has the required shoes and the hat. They are assembled into a parcel and forwarded as freight. The upper order includes just one hat. Beforehand, the warehouse received this hat – together with a second hat for stock – from the supplier and informed inventory management. The warehouse then puts the hat into a parcel, and forwards it as freight.

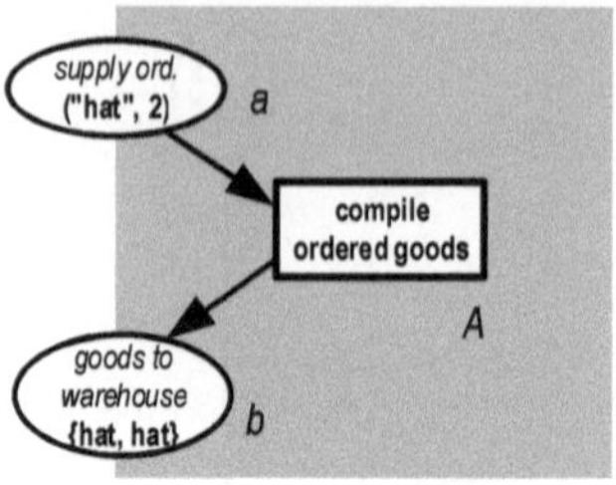

Fig. 11.18 A run of the module *supplier*.

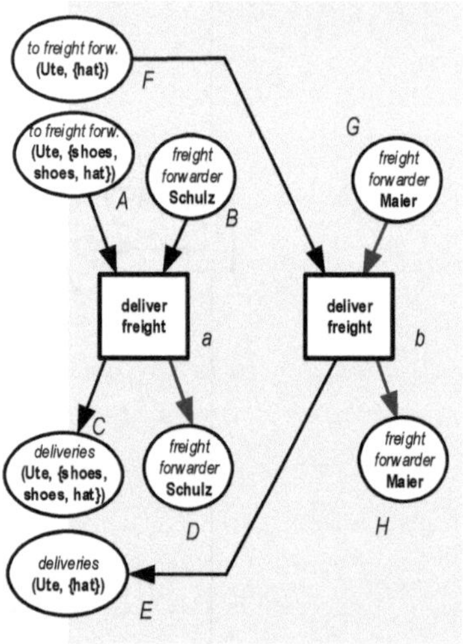

Fig. 11.19 A run of the module *freight forwarders*. The two parcels are delivered to the customer, *Ute*, by different freight forwarders, *Schulz* and *Maier*. No ordering on the two partial deliveries is specified.

may occur side by side with no ordering. This partial ordering is clearly illustrated by the arrangement of the nodes of the overall run in Figure 11.21: Each arrow runs from left to right, but now parts of individual modules are no longer arranged close together. The colored background shows the contributing module. The flow of some modules is divided into several parts. Occasionally the left/right orientation of interface elements of the modules is swapped.

11.5 Dynamics: predicates and event schemata

In the prior part we presented *one* particular run, an example of the behavior of a retailer, where the customer *Ute* sends an order with the article list $[p1, p2, p2]$. This is now to be put in more general terms: there are infinitely many possible article lists for an order from Ute; besides Ute, Max can also submit purchase orders; at the beginning, the three articles, namely, hats, shoes, and pants can be available in the warehouse in different quantities, and order management has many different possibilities to put together parcels. Thus, there are infinitely many possible runs.

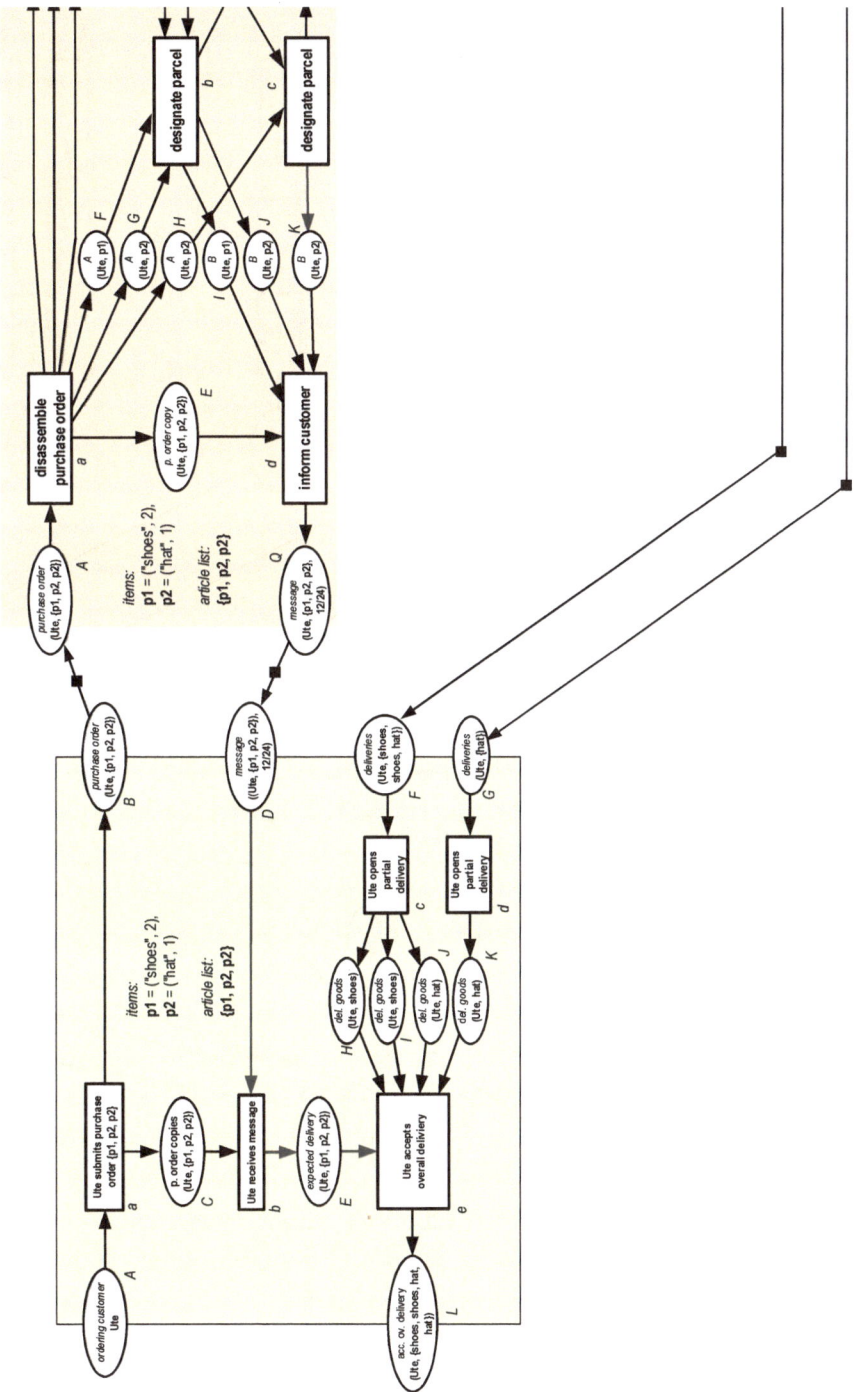

Fig. 11.20 Composition of the runs of the modules: a run of the composed modules (part 1 of 3).

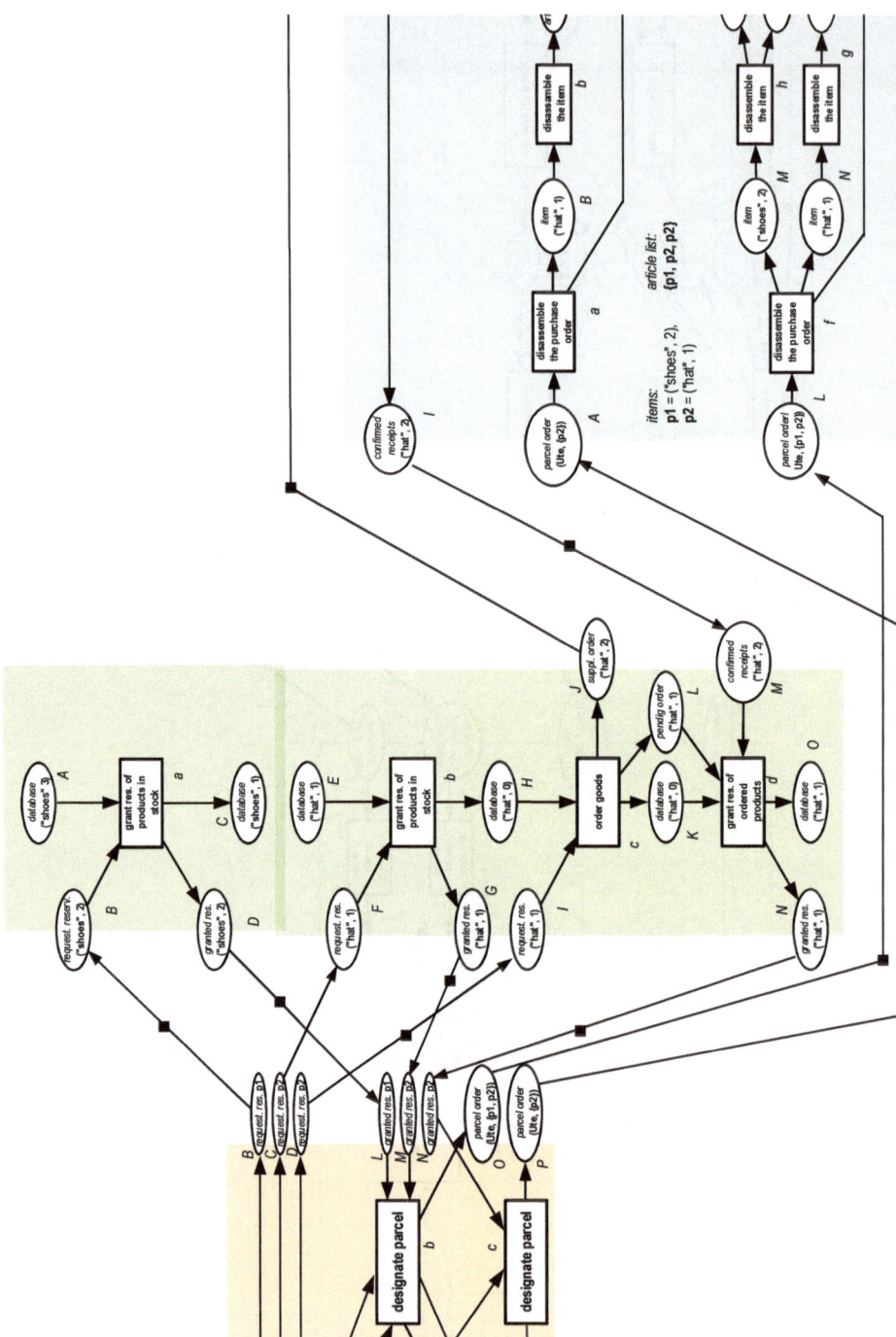

Fig. 11.20 Composition of the runs of the modules: a run of the composed modules (part 2 of 3).

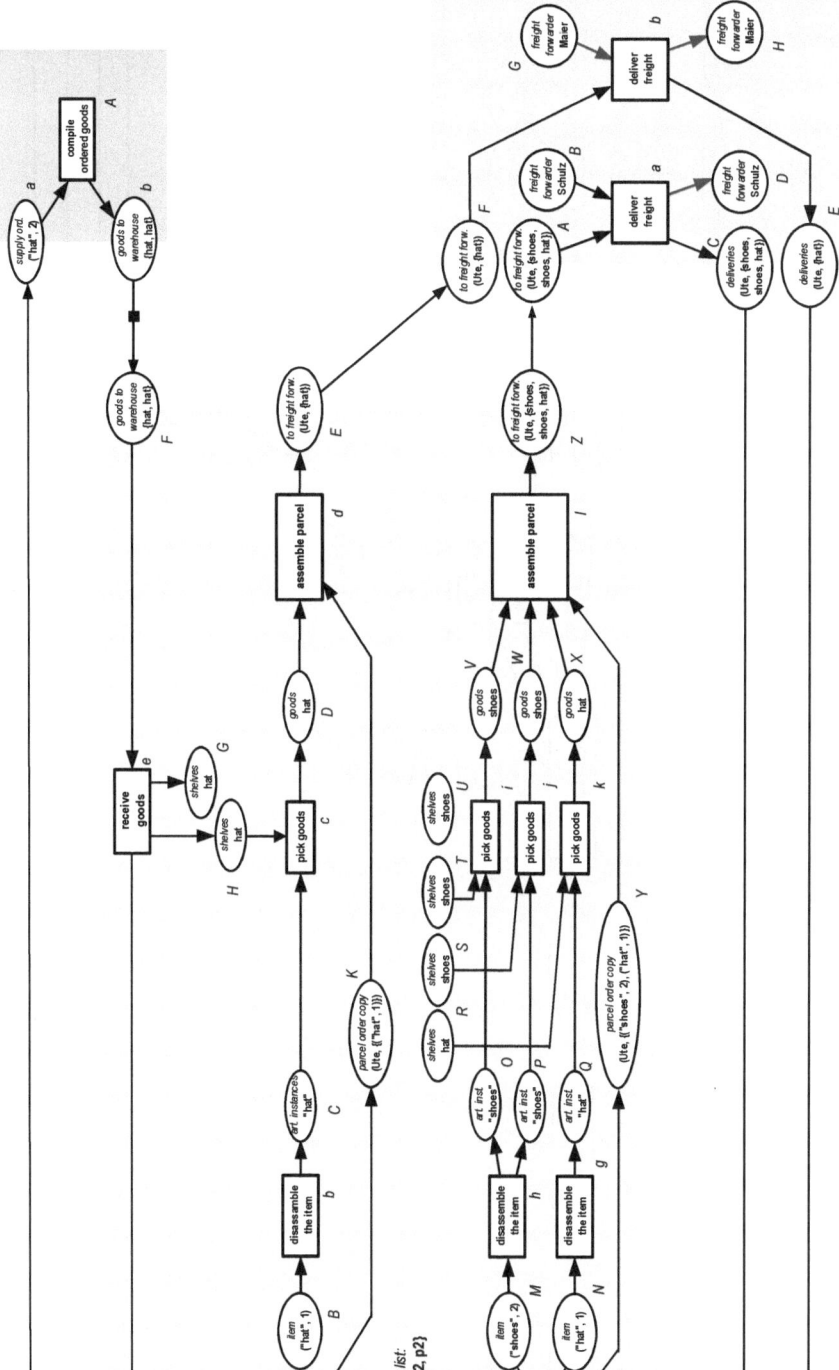

Fig. 11.20 Composition of the runs of the modules: a run of the composed modules (part 3 of 3).

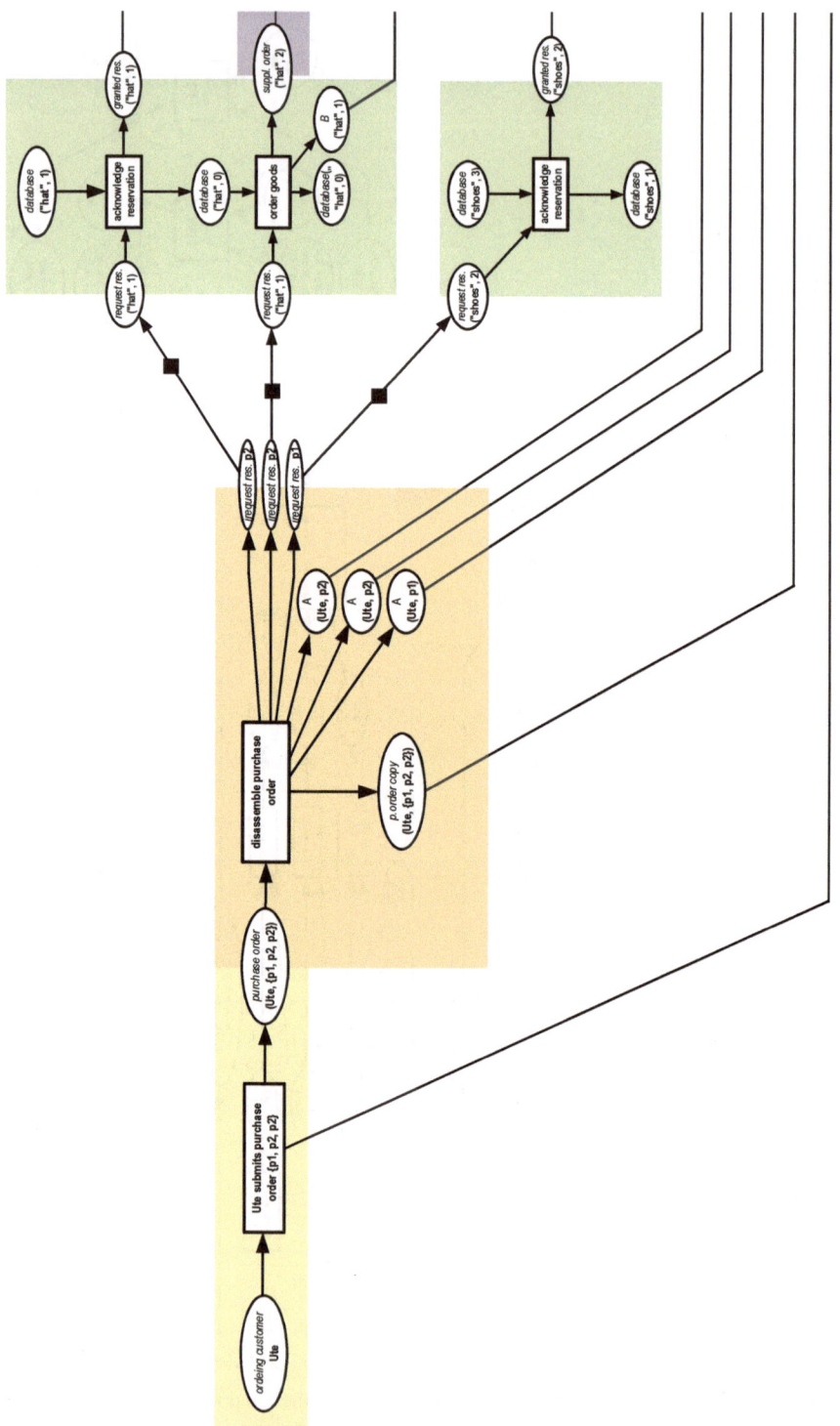

Fig. 11.21 The run, shown from left to right (part 1 of 4).

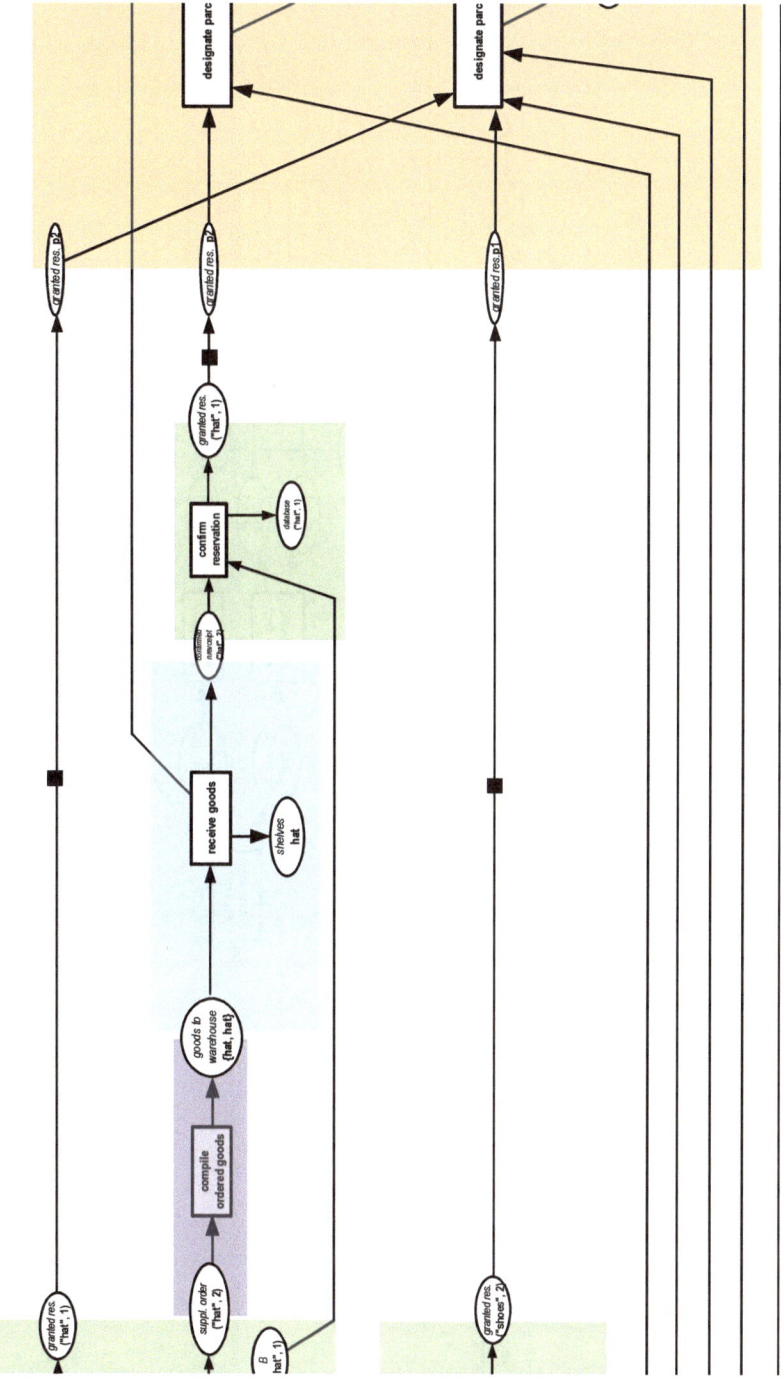

Fig. 11.21 The run, shown from left to right (part 2 of 4).

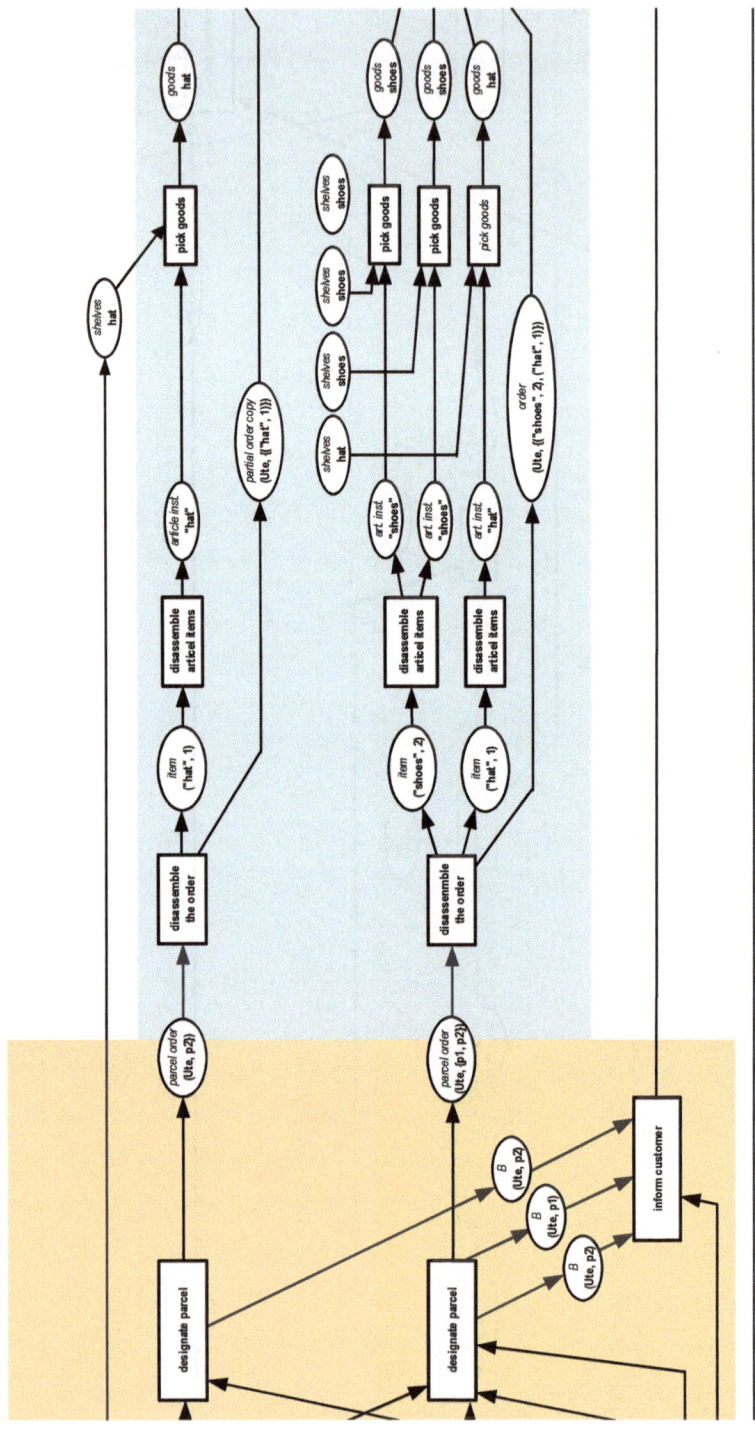

Fig. 11.21 The run, shown from left to right (part 3 of 4).

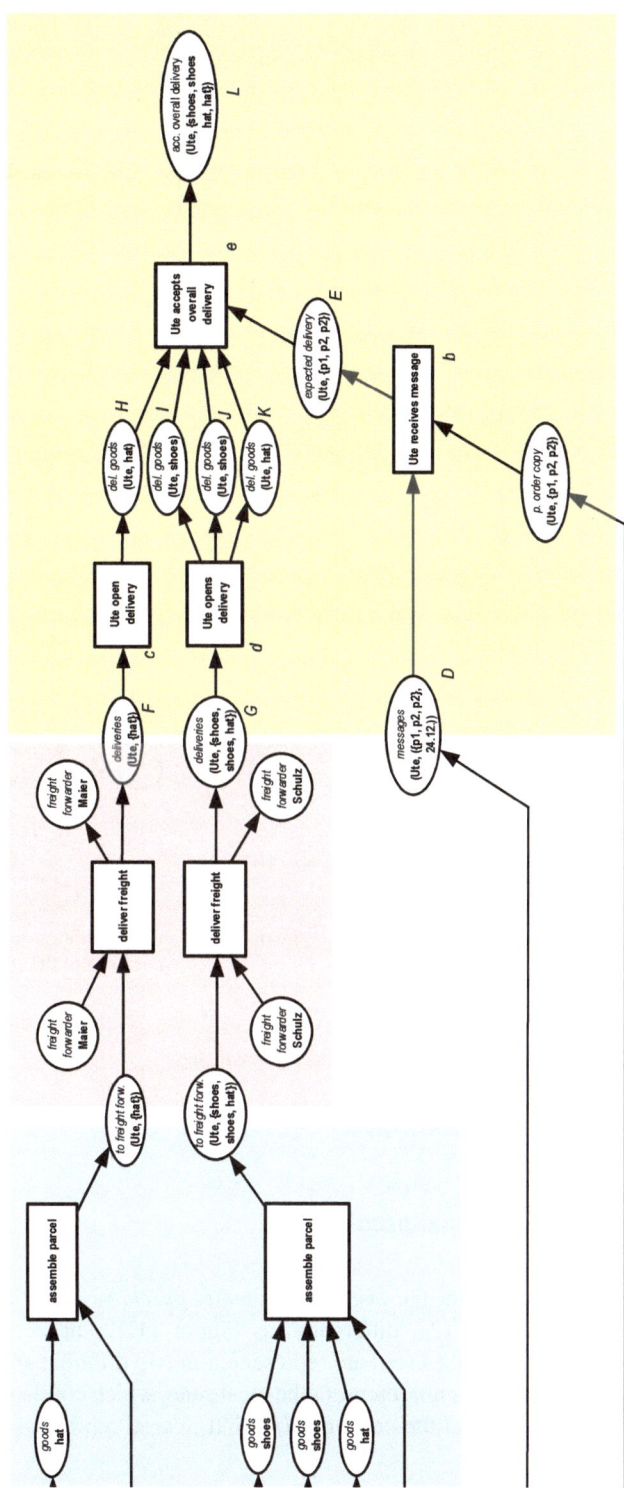

Fig. 11.21 The run, shown from left to right (part 4 of 4).

All these runs shall now be formulated in *one* representation. The idea here is to combine events with the same predicates.

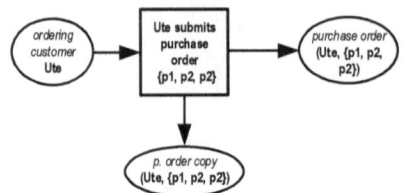

(a) repetition of Figure 11.11, page 127: illustration of an event with statements

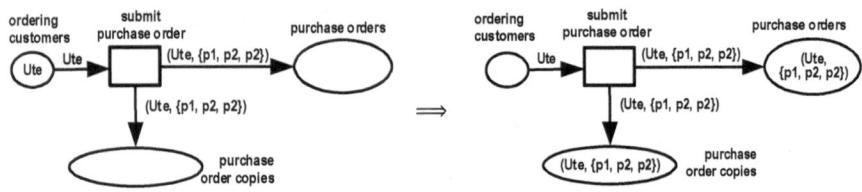

(b) illustration with predicates: before the occurrence of the event

(c) illustration with predicates: after the occurrence of the event

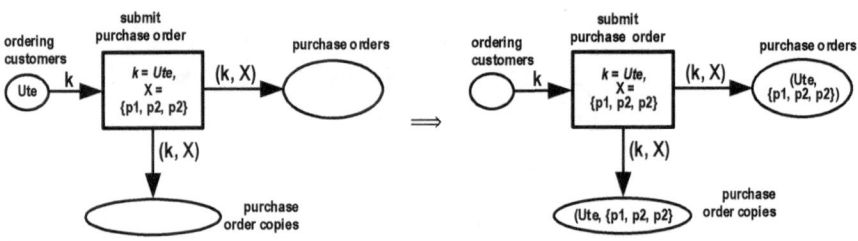

(d) illustration with variables: before the occurrence of the event

(e) illustration with variables: after the occurrence of the event

Fig. 11.22 Transition from local states to predicates.

Events with equal predicates

First, we represent the event *Ute submits purchase order* $\{p1, p2, p2\}$ from Figure 11.11, page 127, differently: as Figure 11.22 shows, the two configurations before and after the event are represented in two different graphs. In Figure 11.22b, the left place does not represent the local state, which consists of the predicate *ordering customers* and the customer *Ute*, but instead represents the predicate *ordering customers*. That this predicate applies to *Ute* is then shown by the token *Ute* within the ellipse. Accordingly, the predicates *sent orders* and *purchase order copies* show that both predicates do not apply to any objects before the event. Figure 11.22c shows the situation after occurrence: the predicate *ordering customers* now does not apply to any object; the predicates *sent orders* and *purchase order copies* now both apply to the object (*Ute*, {*p*1, *p*2, *p*2}). From Figure 11.22b alone, one can derive Figure 11.22c: Intuitively formulated, the label on each arrow determines which objects *flow through* the arrow when the event occurs. Thus, Figure 11.22b represents the same behavior as Figure 11.22a.

In a further step, we replace the arrow labels with variables (Figures 11.22d and 11.22e, page 140), and explain the assignment of the variables in a label of the transition. The representation in Figure 11.22d also represents the behavior of Figure 11.22a.

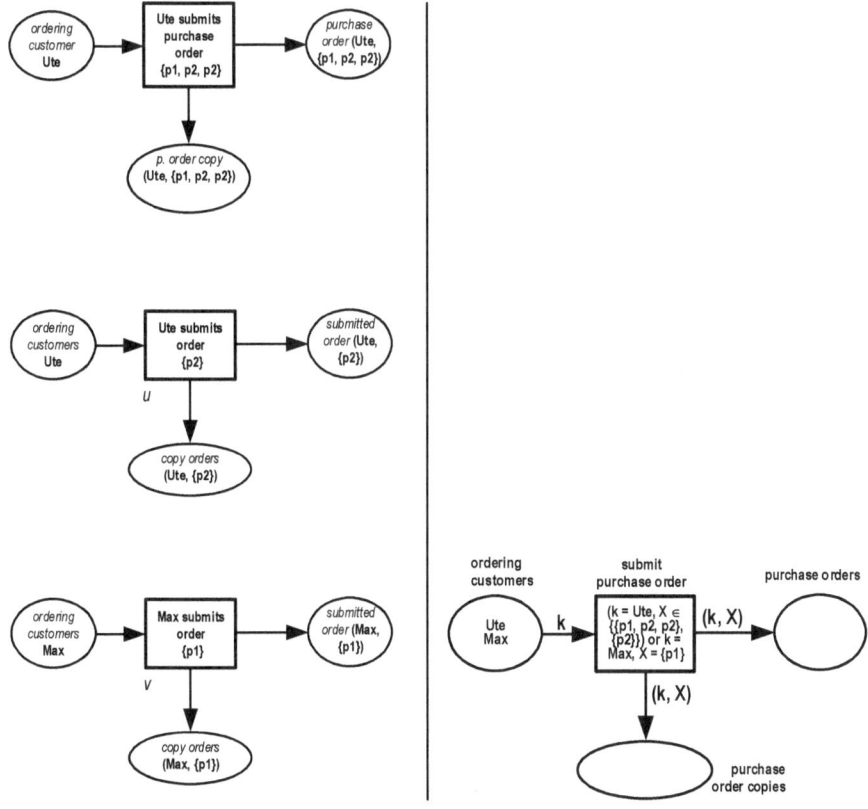

(a) three events represented with statements

(b) common representation of the three events with variables

Fig. 11.23 Representation of events using variables.

Figure 11.23 illustrates the advantage of the new representation. Sub-figure 11.23b shows three different purchase orders: two from Ute and one from Max. Sub-figure 11.23b represents these three purchase orders in *one* schema. The core of the representation is the *variables k* and *X*. They can be *assigned* concrete objects: *k* a customer (*Ute* or *Max*), *X* an article list. Three of these assignments fulfill the label of the *t* transition. Each of them represents one of the three events in Figure 11.23a.

Thus, variables and conditions in the transition *t* can be used to characterize a set of purchase orders. Particularly simple is the characterization of *all* orders which are possible in the given structure: for this purpose any additional condition is omitted.

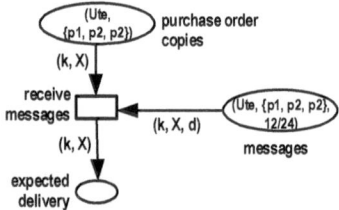

Fig. 11.24 Activated transition *receive messages*.

Figure 11.24 shows how customers deal with messages from the retailer. The delivery date of the last partial delivery for an order is communicated to the customer: The transition occurs when the variables k and X can be assigned to a customer k_0 and an order list X_0 in such a way that (k_0, X_0) is a token lying on the place *purchase order copies* and for any date d_0 the token (k_0, X_0, d_0) lies on the place *messages*. In Figure 11.24, the variables, k, X, and d are assigned tokens Ute, (p1, p2, p2) and 12/24.

Functions

Figure 11.25a repeats Figure 11.18: The supplier receives the item ("hat", 2) and delivers the multi-set [*hat, hat*]. We need a general principle to derive a multi-set of goods from an article position. For this we use a function that assigns an article to each good. In our case study this is the function f with $f(hat) = $ "hat" and $f(shoes) = $ "shoes". In general, f is not injective; for example, a car rental company could offer small, medium, and large vehicles, three types of goods, but has many real vehicles. Each one is small, medium, or large.

To schematically characterize the supplier event, Figure 11.25b uses the variable w for the goods to be delivered and the variable p for their number. The single-element multi-set $[w]$ with the factor p yields the multi-set $p \cdot [w]$, which contains p instances of the goods w. The function $\_'$ returns the corresponding articles for an *item* or *article list*.

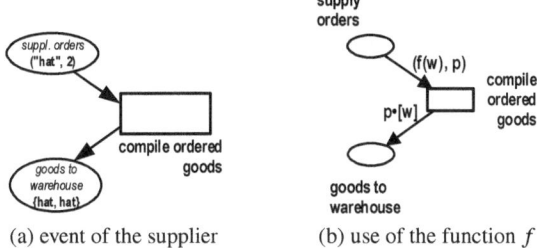

(a) event of the supplier   (b) use of the function f

Fig. 11.25 System functions.

Loops

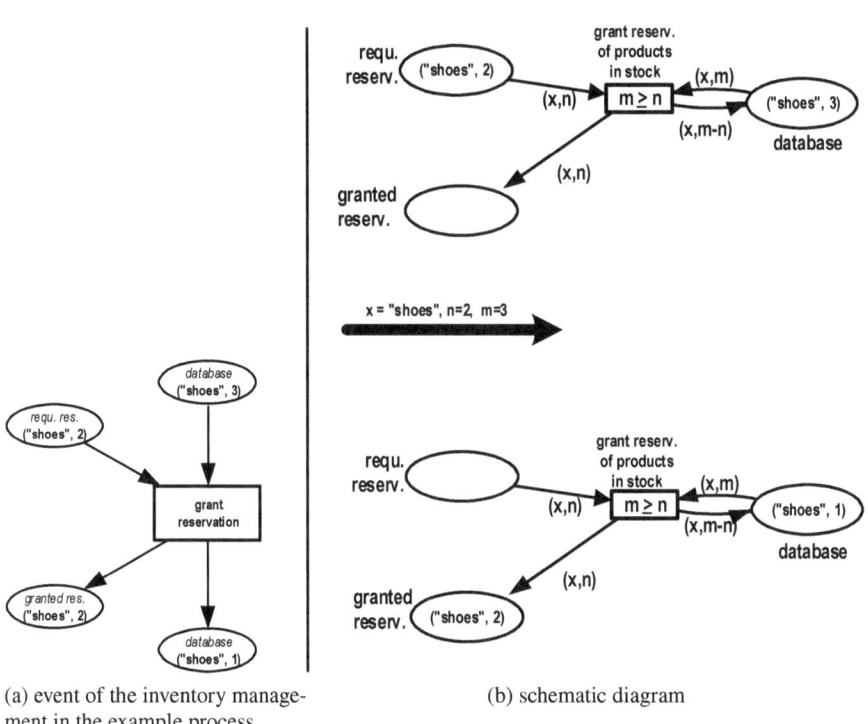

(a) event of the inventory manage-   (b) schematic diagram
ment in the example process

Fig. 11.26 Event and associated schematic representation with a loop.

It can happen that during an event an object leaves a local state and another
(or the same) object reaches a local state with the same predicate. An example
of such a predicate is the inventory management database, which is shown again

in Figure 11.26a. In a schematic representation, each predicate occurs only once. Therefore, in Figure 11.26b a loop is created between the space *database* and the transition *confirm availability*.

elm-notation

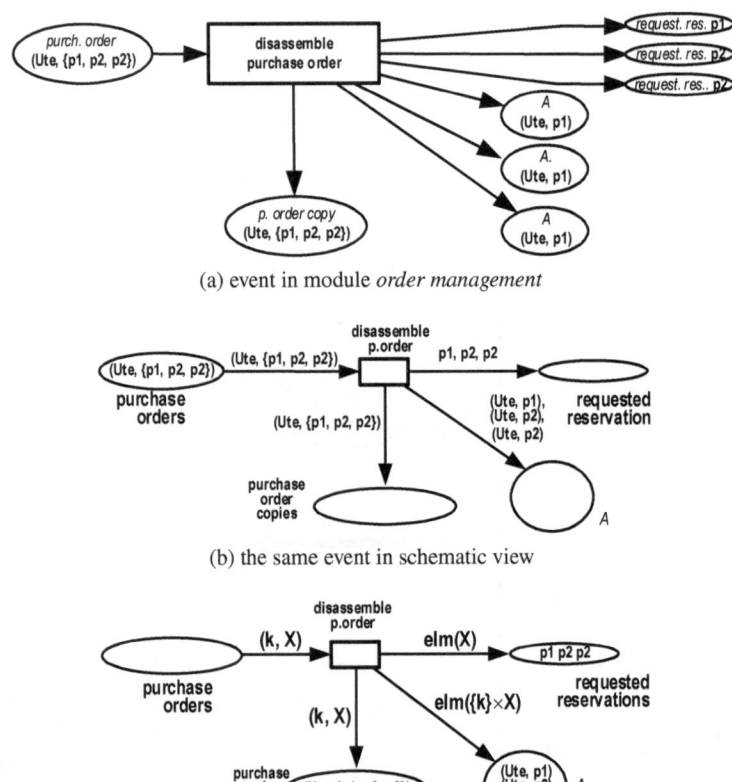

(a) event in module *order management*

(b) the same event in schematic view

(c) the same event with variables in mode $k = Ute$, $X = \{p1, p2, p3\}$

Fig. 11.27 The use of *elm*.

In the previous examples of schematic representations of events, at most *one* object leaves or reaches a place. This is not always the case. For example, the event *decompose order* in the module *order management* generates three local states with the predicate *requested reservation* (Figure 11.27a). Therefore, in the schematic representation (Figure 11.27b) the three objects $p1, p2, p2$ are located on the arrow towards the place *requested reservation*. When *disassemble purchase order* (abbreviated as *disassemble p. order*) occurs, the three objects reach this place at the same time. When using the variable X with the assignment $X = [p1, p2, p2]$, the arrow label X would place the set $[p1, p2, p2]$ as one token on the place. However, we expect the three tokens $p1, p2, p2$, that is, the elements of this set (Figure 11.27c). The notation $elm(X)$ ensures that after *decompose order* the elements of the set X reach the place *requested reservations*, and not the set itself. The same applies to the place *copy position* and the arrow ending there.

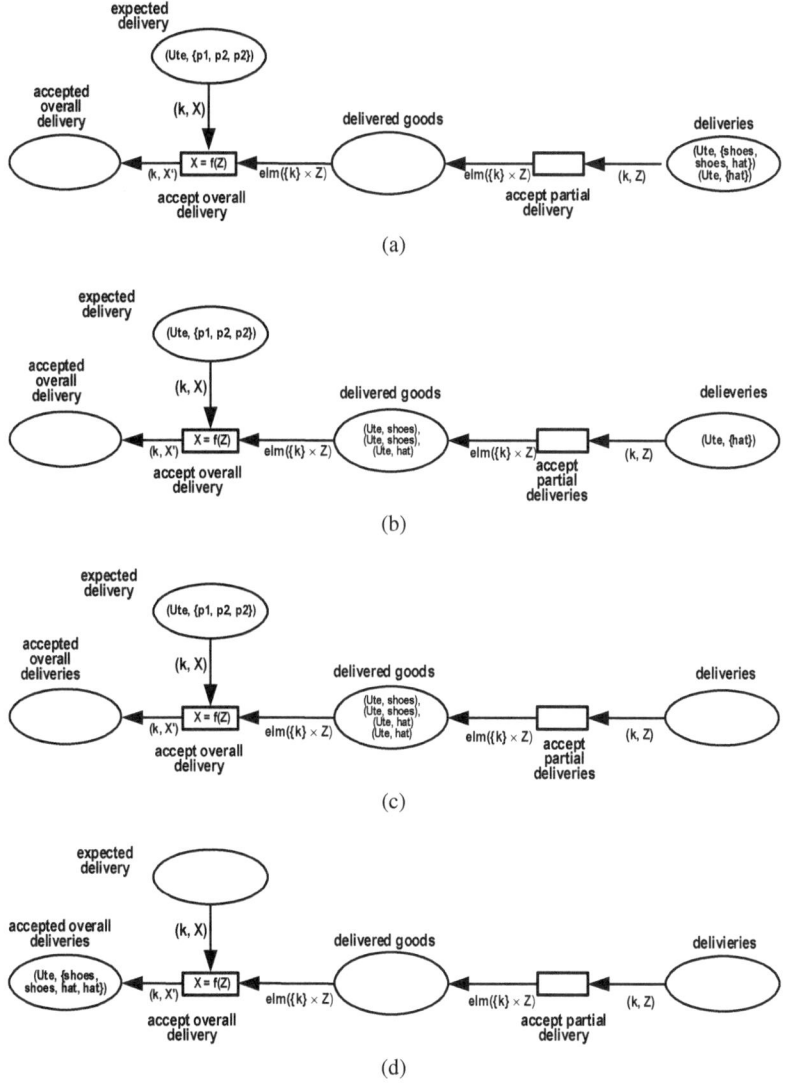

Fig. 11.28 Receive goods.

As another example, Figure 11.28 shows how partial deliveries are compared with the order and accepted as a complete delivery: In Figure 11.28a we assume two partial deliveries. First, an incoming delivery is opened and each of the goods is individually marked with the customer *Ute* (Figure 11.28b). After the second partial delivery, there are four goods on the place *delivered goods* (Figure 11.28c). The quantity of these goods is assigned to the variable Z; $f(Z)$ indicates the quantity of the ordered articles. This quantity is derived with X' from the article list $X = [p1, p2, p2]$. With this, the condition in the transition is fulfilled, the transition occurs, and the complete delivery is accepted (Figure 11.28d).

11.6 Integrating architecture, dynamics, and statics: The retailer for the structure S

We now have all the means of expression at our disposal to model all aspects of the case study. In the logic of an order and its processing, we start with the customer model. Then we follow with the three departments of the retailer (order management, inventory management, and warehouse) and finally the two other business network partners: the supplier and the freight forwarders. The models are based on structure S, which was presented in Figure 11.10, page 126.

The customer model

Figure 11.29 shows the module *customers*. With the set $K = \{Ute, Max\}$ from the structure S, the two tokens *Ute* and *Max* are located at the beginning in the place *ordering customers*. We have already discussed the transitions in detail. In the context of the other modules, Figure 11.30 shows a reachable marking from which Ute receives a message with the delivery date (12/24) and then receives and accepts the entire delivery in two partial deliveries, as discussed in Figure 11.28. Independent of this, Max can submit a purchase order.

The order management model

We have already discussed the transition *disassemble purchase order* of the module *order management* (Figure 11.31) in Figure 11.27. The order management asks for a confirmation of the availability of each individual order item. This can be delayed if, for example, some units have to be reordered from the supplier first. As soon as some order items have reached the place *granted reservations*, the order management assigns the warehouse a corresponding partial delivery. The order items sent in partial deliveries are collected on place B; when all items of an order are assigned to

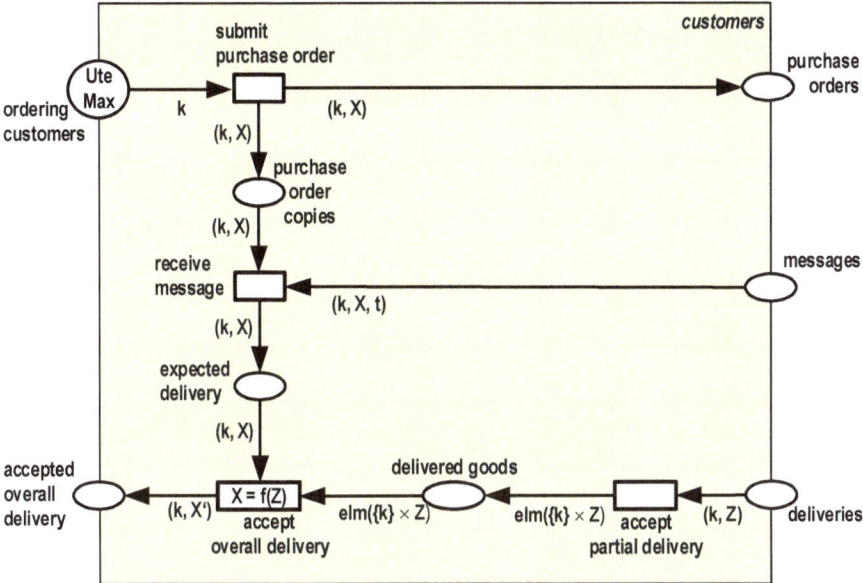

Fig. 11.29 The module *customers*. (1) Place *ordering customers*: This place initially contains a token for each potential customer. A purchase order consists of a customer k, and an article list X. (2) Transition *receive message*: This transition occurs if the received message is identical to a copy of a previously submitted order, extended by a date. (3) Transition *accept partial delivery*: A partial delivery Z for the customer k is opened, and each single good in Z turns into a token of its own. There may arrive many such partial deliveries for the same customer, k. (4) Transition *accept overall delivery*: This transition occurs when, for each item on the article list on the place *expected delivery*, there is a corresponding number of goods on place *delivered goods*.

partial deliveries, the transition *inform customers* sends a message to the customer with the scheduled date for the last partial delivery of the order. Technically, the order management can assign any date to the variable t.

The inventory management model

In the *inventory management* model in Figure 11.32, the entries of the article list G in place *database* describe, for each ordered article, the number of matching available goods in the warehouse, initially three pairs of shoes, a hat, and no shirts. If there are at least n units available in the warehouse for a requested order item (a, n), transition a releases n units for a partial delivery. If not, transition b orders the n units of the article a from the supplier as well as a stock of a further p units. By defining p as a variable, inventory management can freely decide the amount of stock for item a for each reorder. If p were declared as a constant, the S structure would set the stock level for the item a once and for all.

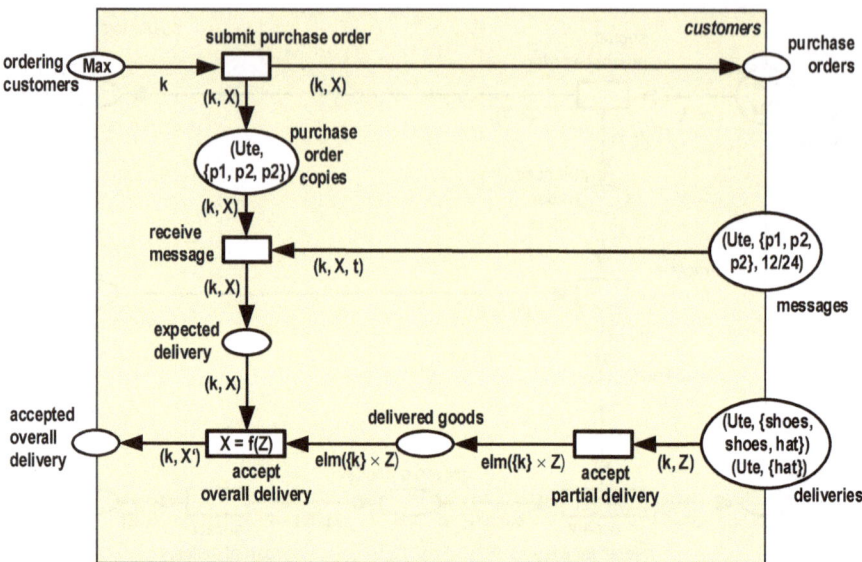

Fig. 11.30 A reachable marking of the *customers* module in the context of all modules. (1) After having submitted a purchase order, Ute eventually receives a message and two partial deliveries arrive; tokens are on place *deliveries*. (2) Both deliveries are opened. Together they contain two pairs of shoes and two hats. This corresponds to the purchase order and is accepted as the overall delivery.

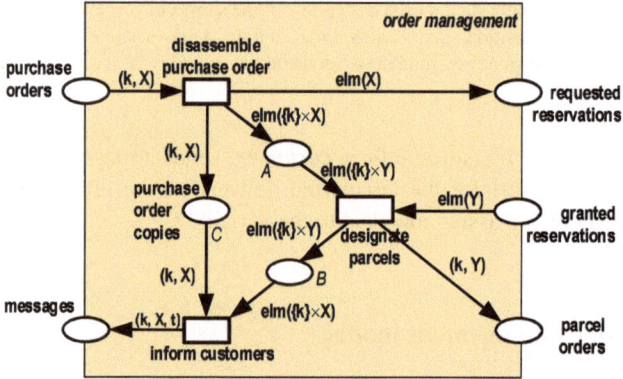

Fig. 11.31 The module *order management* queries availability of goods and orders partial deliveries. (1) Transition *disassemble orders*: The order management disassembles the article list X of an incoming order (k, X) of a customer k into its items; each of these items is placed on place *requested reservations*, as well as on place A. Each item in A is additionally indexed with the customer k. (2) Transition *inform customers*: Place B collects the items of partial deliveries for customer k. When the items complete the purchase order, the customer is informed about the date t of the last partial delivery. (3) Of particular interest is the transition *designate parcels*: Order management selects a subset Y of items from all granted reservations. Then a parcel for Y is designated, and all items in Y are collected in place B.

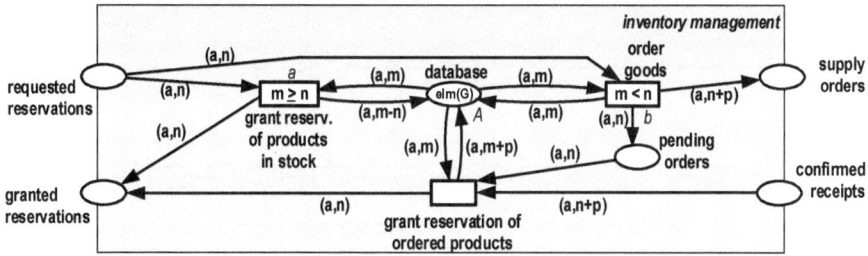

Fig. 11.32 Inventory management confirms the availability of article. (1) Place *database* reports the goods in the warehouse. This list is a set G of items, indicating for each good the corresponding number of available instances. Each such item is a token on the place *database*. (2) A request for the reservation of an item (a, n), i.e. for n instances of an article a, has two potential outcomes: (i) The number m of available instances meets or exceeds the number n of requested instances. In this case, the transition *grant reservation of products in stock* grants reservation of the goods, and reduces the corresponding number in the corresponding item of *database*. (ii) Otherwise, inventory management orders $n+p$ instances from the supplier, increasing the number n of requested instances by p, as stock for forthcoming inquiries. The inventory management may select a different number p at each occurrence of *order goods*. (3) The warehouse will eventually acknowledge receipt of $n + p$ instances. Then, the inventory management updates the corresponding item in *database*, and acknowledges reservation of n instances to order management.

When the request to restock an item triggers an order to the supplier, the request is kept on place *pending orders* until the warehouse has confirmed receipt of the units from the supplier. The transition *grant reservation of ordered products* then passes this confirmation to the order management and updates the information about the stock in place *database*. Place *database* thus describes for each item how many units have not yet been reserved for delivery in the warehouse. Here, one can see a subtle difference between one large order quantity and several small order quantities of a particular item: If, for example, three shirts are available and one order requests four shirts, that whole request is held back on place *pending orders* until the supplier has delivered more shirts. If, instead, two orders requesting two shirts each are made, one of them is confirmed immediately and in a partial delivery the customer can already receive the first two shirts before the supplier delivers the other shirts to the warehouse.

The model of the warehouse

The warehouse in Figure 11.33 has two tasks: for each incoming order it packs a freight parcels for the freight forwarders; furthermore, the warehouse processes the incoming deliveries from the supplier. For the first task, the transition *disassemble the order* breaks down an incoming article list into its items. From the item (a, n) of an article a, which is ordered in quantity n, the transition *disassemble articles* then creates n tokens of type a: first, $n \cdot [a]$ creates the multi-set containing n instances of type a. With $elm(n \cdot [a])$, each of its n elements then becomes an individual

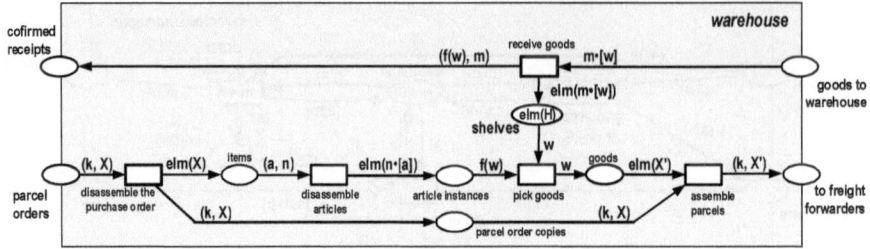

Fig. 11.33 Warehouse packs freight parcels and confirms the goods sent by the supplier. This module performs two tasks: (1) According to the orders coming from *order management*, the module assembles parcels and forwards them to the *freight forwarders*. (2) The module receives incoming goods from the *supplier* and informs *inventory management*. *Ad (1)*: The module stores each incoming order in place *parcel order copies*, and additionally disassembles the order into its *articles*. Then, transition *disassemble article items* generates, for each item (a, n) of an article, n tokens of the form "a" (instances of a) on the place *article instances*. For each item, transition *pick goods* picks corresponding goods from the shelves and forwards them to place *goods*. The function f describes the goods w that would correspond to the article $f(w)$. Transition *assemble parcels* bundles the goods according to the article list X and forwards the parcel to the freight forwarders. *Ad (2)*: The transition *goods to warehouse* loads the instances of an incoming goods delivery onto the *shelves*, and informs *inventory management*.

token. For each of these tokens a, the transition *pick goods* takes a good w from the warehouse which corresponds to a, to which therefore $f(w) = a$. This commodity is put on place *goods*.

For the second task the transition *receive goods* receives a parcel $m \bullet [w]$ from the supplier with m instances of the good w. The instances are individually stored in the warehouse ($elm(m \bullet [w])$ is used to create m tokens of the type w). At the same time, inventory management is informed about the receipt of the goods: $f(w)$ describes the article of the goods w, $(f(w), m)$ also the quantity.

Place *shelves* represents the actual stock; it initially contains the elements of the multi-set H, that is three pairs of shoes, a hat, and no shirts. This corresponds to the initial marking $elm(G)$ of place *database* of the module *inventory management*.

The supplier model

Figure 11.34 shows how the *supplier* receives a supplier order and selects an item w that matches the ordered item $f(w)$. The *supplier* then transports a package of p instances of this item to the goods receiving area of the warehouse (token $p \bullet [w]$). In this model, the *supplier* can always deliver; of course, it is possible to model other assumptions here.

Fig. 11.34 The module *supplier* sends corresponding goods for each order. (1) Transition *compile ordered goods*: From the *inventory management*, the *supplier* receives the order of p instances of article a, and delivers p instances of "matching" goods w. A good w "matches" an article a if $f(w) = a$. (2) Different goods may match the same article: the goods may be shaped differently or may have different control numbers, et cetera.

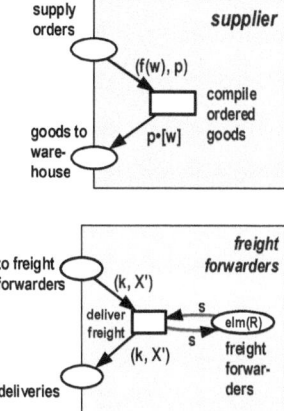

Fig. 11.35 The module *freight forwarder* delivers freight parcels. Transition *deliver freight* selects a freight forwarder, who accepts freight from the warehouse, and delivers it to the customer.

The model of the freight forwarders

The model of the freight forwarders in Figure 11.35 is obvious: There is a basic type R of freight forwarders. Initially, the place *freight forwarders* contains some freight forwarders. The variable s is used to select a freight forwarder. This module provides a starting point for further assumptions regarding the selection of freight forwarders.

All modules of the case study are now modeled. The models can be composed as shown in Figure 11.36. This creates the model of the retailer for the structure S.

11.7 The overall case study model

In the model shown in Figure 11.36 the basic domains and basic functions, which are the customers, goods, initially available goods, assignment of articles to goods, and the freight forwarders, are fixed. A model that includes all possible basic domains and basic functions would be interesting. Furthermore, the model itself should be purely *symbolic*, so it should not work with specific quantities and functions like the models in Section 11.5, page 132, but with symbols for quantities and functions. Then the model could be processed with software tools.

Here we can again take up proven concepts of predicate logic and many specification languages: We construct a *signature* Σ for the retailer. The basic idea is simple: Σ contains a symbol for each basic domain, each derived domain, each constant, and

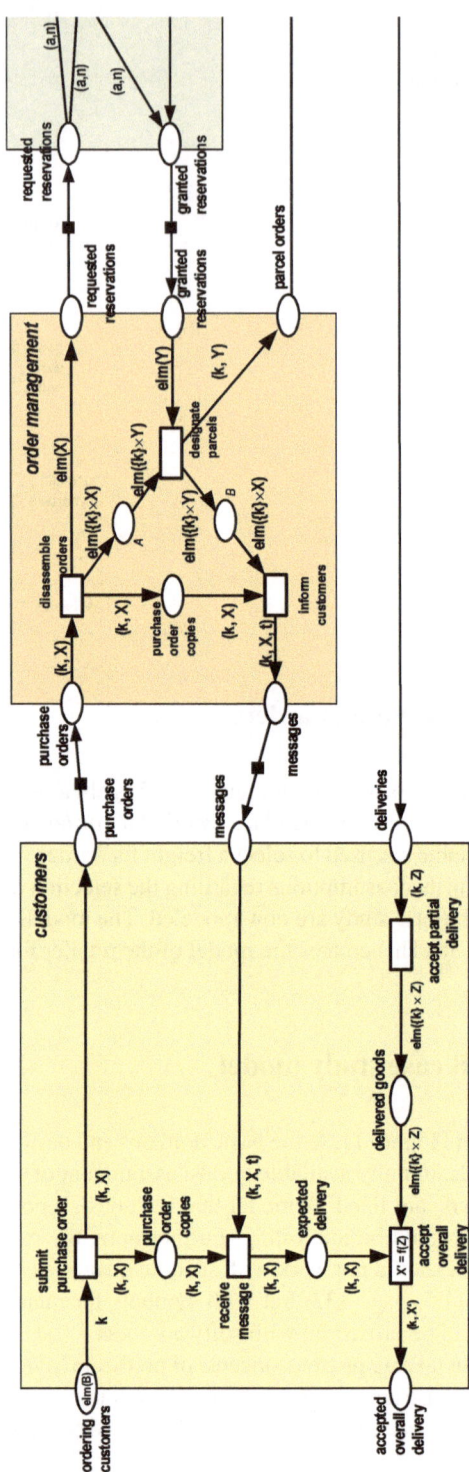

Fig. 11.36 Model of the retailer, composed of six modules (part 1 of 3).

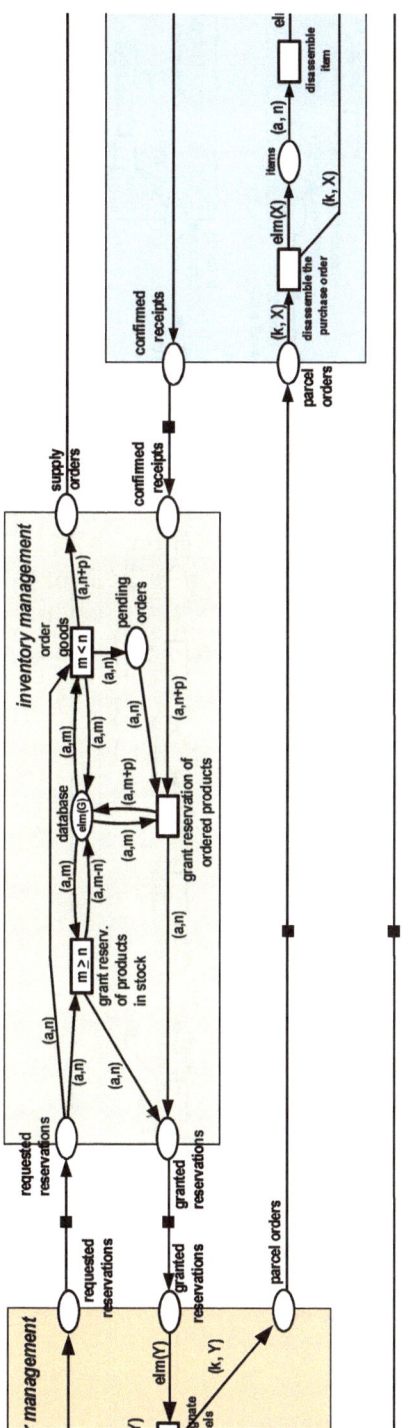

Fig. 11.36 Model of the retailer, composed of six modules (part 2 of 3).

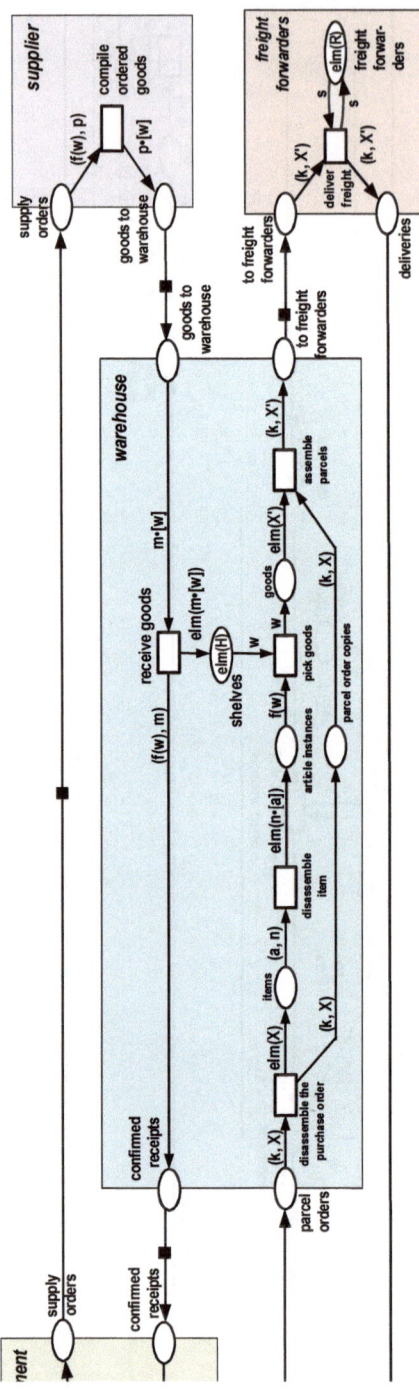

Fig. 11.36 Model of the retailer, composed of six modules (part 3 of 3).

basic symbols	constant symbols	properties
KN *customers*	p1, p2: AP *ordered article positions*	$(a,n)' = n[a]$ for $(a,n) \in$ AP
AR *articles*	B: P(KN) *ordering customers*	$[p1, ... , pn]' = p1' + ... + pn'$ for
WA *goods*	G : AL *initially listed articles*	$[p1, ... , pn] \in$ AL
TE *dates*	H : WM *initially available goods*	
SP *freight forwarders*	R : P(SP) *available freight forwarders*	$G' = f(H)$

derived symbols	function symbols	variables
AP = AR × N *items*	f : WA $\longrightarrow$ AR	k: KN x: AR X,Y: AL Z: WM
AL = *M*(AP) *article lists*	f : WM $\longrightarrow$ AM	t: TE w: WA s: SP m, n, p: N
AM = *M*(AR) *sets of articles*	(_'): AP $\longrightarrow$ AM	
WM = *M*(WA) *sets of goods*	(_'): AL $\longrightarrow$ AM	

Fig. 11.37 Signature Σ.

each function. An *instantiation* of the signature assigns a matching set or function to each such symbol. Figure 11.37 shows such a signature. An *instantiation* then assigns a set, object, or function to each symbol according to its type. The structure S is an instantiation of the signature Σ.

To describe the behavior of instantiations, we can use the nets from Section 11.6, page 146, as they are. This is because these nets do not use specific aspects of the instantiation S; for example, the initial marking of the place *ordering customer* is not labeled with *Ute* and *Max*, but $elm(K)$. The instantiation S then results in Ute and Max as the initial marking of the place *ordering customers*. Somewhat more difficult is the general formulation of the relationship between stored positions of available items in inventory management and the available goods in the warehouse: The function f, which can be freely chosen, describes which different goods fulfill an ordered item request.

11.8 Further perspectives on the retailer

Abstract environments for behavioral models

There are a number of other perspectives on the retailer and its business network partners. For the behavioral module of each partner there is an abstract module, its environment, so that the composition of the two modules results in the overall module. Figure 11.38 shows the behavior of the retailer's three business network partners in their respective environments.

Module for orders and flow of goods

Figure 11.36 composes the six modules involved in the retailer and thus describes the behavior of the entire system in a structured way: The components remain recog-

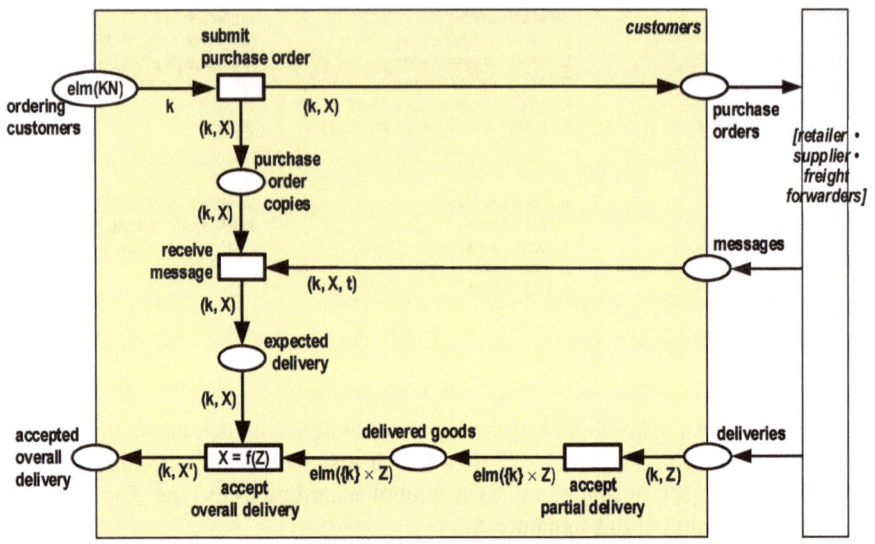

(a) perspective of the customer

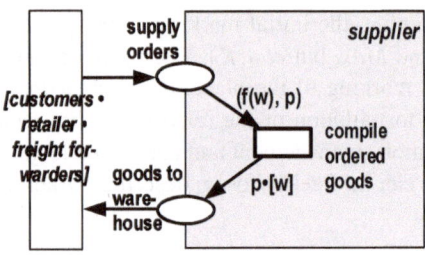

(b) perspective of the supplier

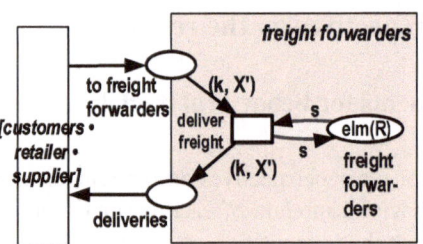

(c) perspective of the freight forwarder

Fig. 11.38 Perspectives of the business network partners on their respective environment.

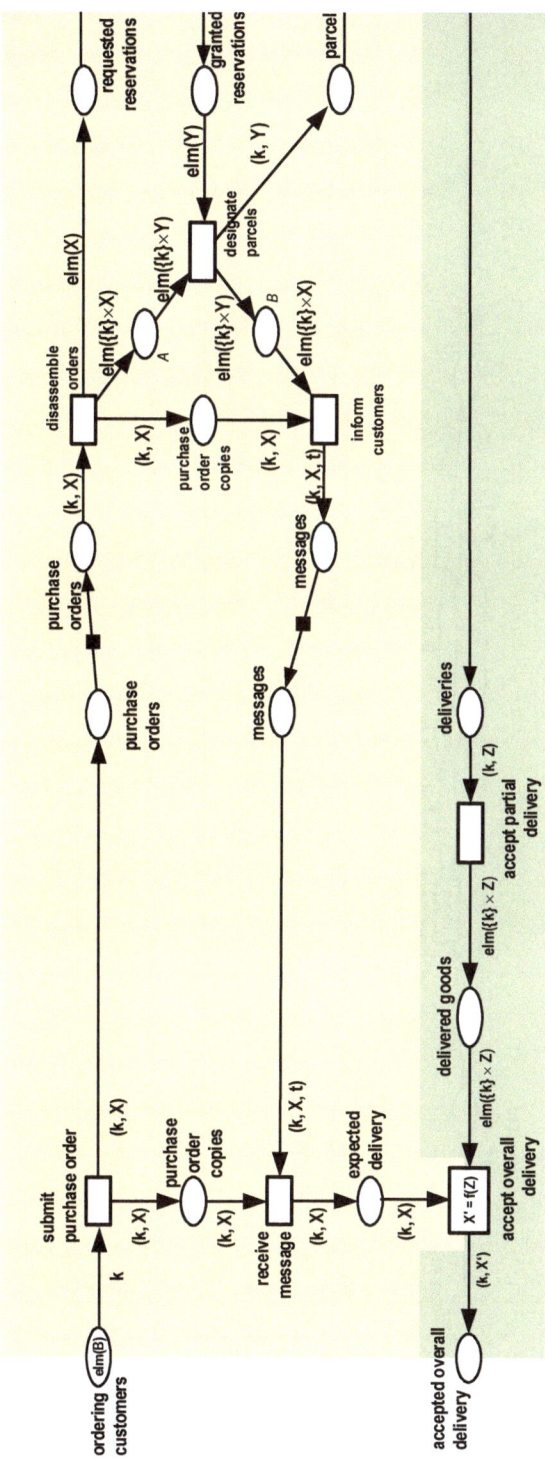

Fig. 11.39 Modules of orders (yellow) and flow of goods (green, part 1 of 3).

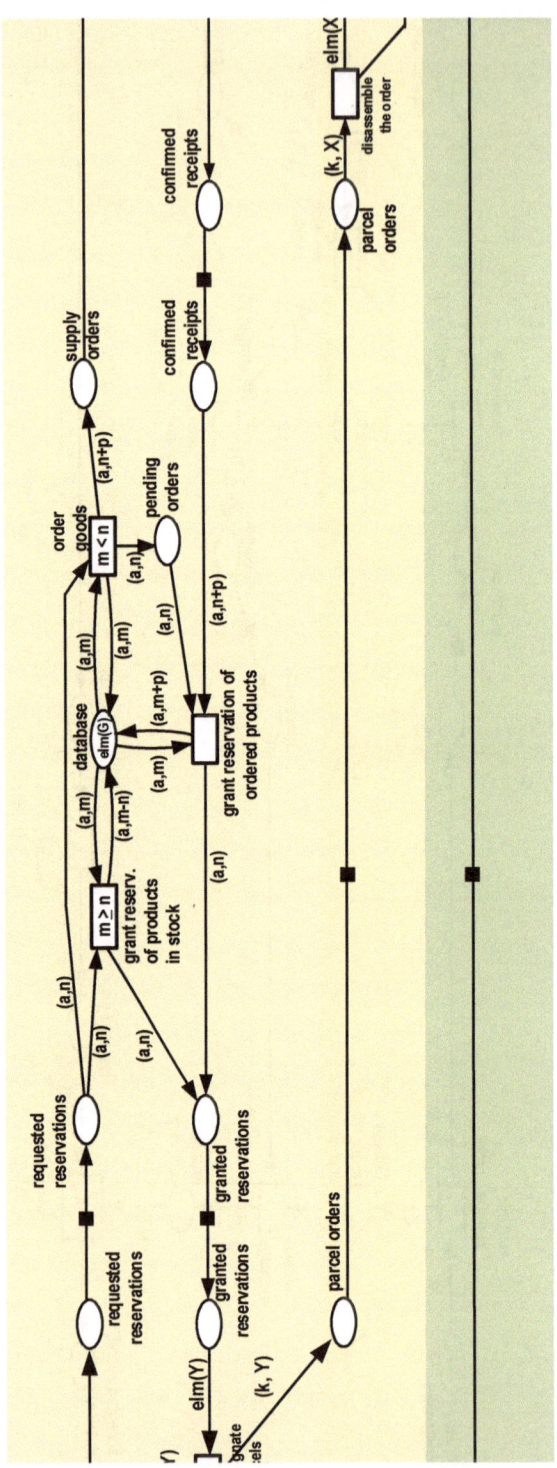

Fig. 11.39 Modules of orders (yellow) and flow of goods (green, part 2 of 3).

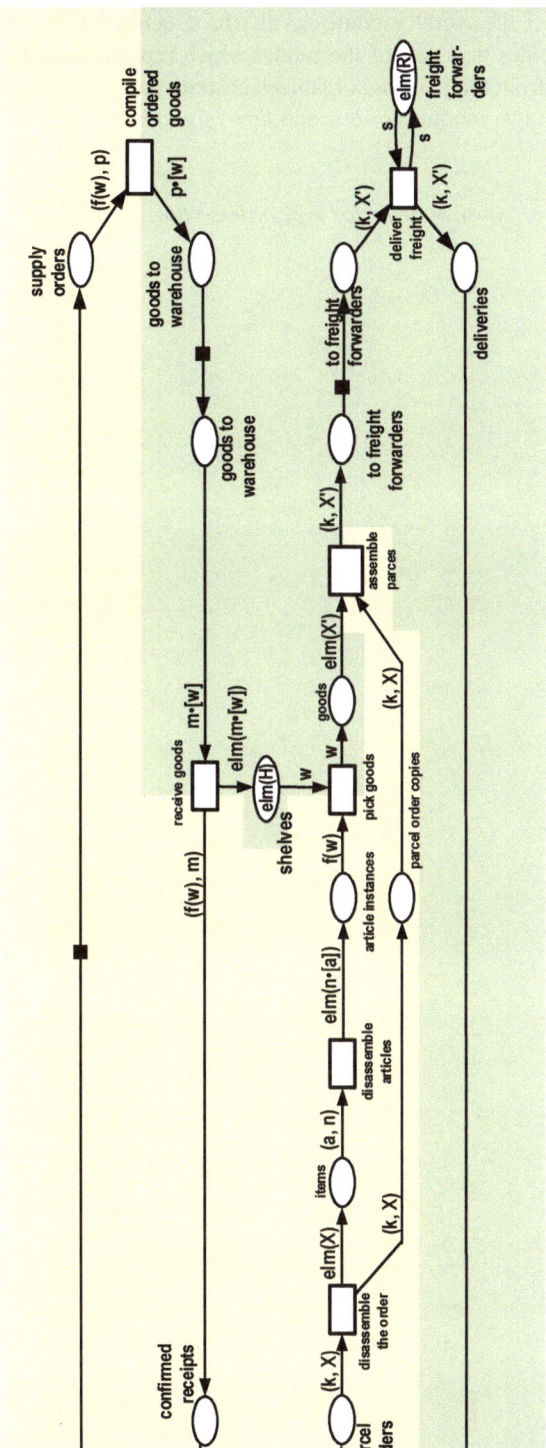

Fig. 11.39 Modules of orders (yellow) and flow of goods (green, part 3 of 3).

nizable. However, the overall module can also be structured differently. Figure 11.39, page 157, separates that part of the model which represents real physical objects, namely goods, from the part that organizes abstract data of the orders. This separation leads to the two modules *orders* and *flow of goods* in Figure 11.39. It therefore holds that:

$$overall\ behavior =_{def} orders \bullet flow\ of\ goods\,. \tag{11.7}$$

Chapter 12
HERAKLIT guidelines

12.1 How to start modeling

The less experienced modeler may wonder how to convert a given informal description of a system into a formal HERAKLIT model, in a more or less systematic manner. Here, we give some rules of thumb for this endeavor, and exemplify them by above-considered as well as new examples.

Some general hints

First of all, the modeler should keep in mind that HERAKLIT bases a system model on three pillars: *architecture*, *dynamics*, and *statics*. Sometimes, especially in the case of a small system, the architecture is not very pronounced. In other cases, items and data are absent or abstracted away, in favor of abstract data- or control-tokens. Behavior, however, is represented in every sensible HERAKLIT model.

Technically, the modeler should always think in terms of *modules*. As outlined in Part I, a module has two interfaces and an inner part. HERAKLIT distinguishes three types of modules, reflecting the interior of the modules:

1. *General* modules: the interior of a general module is not limited in any way; it may consist of (informal) text documents, diagrams, or other objects. One extreme example is software inside a module that sends remote procedure calls (RPC) and receives responses along the module's interfaces. Another example is so-called *boundary objects* or knowledge objects, which are typically discussed in (management) information systems as important links between the real and the formal world.
2. *Abstract* modules: the interior of an abstract module is empty, or is composed of abstract modules.
3. *Behavioral* modules: the interior of a behavioral module is a Petri net. A comprehensive module may contain all three kinds of modules as sub-modules. Con-

© The Author(s), under exclusive license to Springer Nature Switzerland AG 2024
P. Fettke, W. Reisig, *Understanding the Digital World*,
https://doi.org/10.1007/978-3-031-61898-7_12

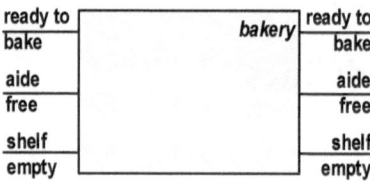

Fig. 12.1 Most abstract view of the bakery.

structing a model of an existing or imagined system usually uses all three versions of modules.

Furthermore, using HERAKLIT as a new digital way of understanding the world can be started on different levels. We distinguish five different levels of maturity in the use of HERAKLIT:

1. Level: This is the starting point, without any specific requirements on the model description. Every project starts at this level.
2. Level: HERAKLIT modules are used to understand and structure the system. The interior of modules may contain informal objects, e.g. texts, pictures, videos, or diagrams of different modeling languages such as BPMN, UML, and event-driven process chain. Module interfaces are carefully chosen and named; though they are subject to later refinement and update.
3. Level: In addition to the requirements of level 2, the interior of some HERAKLIT modules use HERAKLIT concepts.
4. Level: Compared to level 3, the majority of the interior of HERAKLIT modules use HERAKLIT concepts. However, concepts from other modeling frameworks also appear.
5. Level: Only HERAKLIT concepts are used to model the system under consideration.

Two examples of modeling at level 5 are the two case studies reported in this book. Note that we generally recommend understanding the digital world at this level. However, it is obvious that many modeling projects cannot be started from scratch, e.g. often such a "greenfield" approach is not efficient. In such contexts, it is easily possible to use HERAKLIT concepts to structure the overall system and to incorporate the concepts of other frameworks within HERAKLIT.

Starting with architectures

In a systematic way, the modeling of a given or imagined system starts with a most abstract, general module. This module comes with an interface that describes the system's link to its environment, as in Figure 12.1 for the case of the bakery example.

A module's purpose usually implies clearly which elements must go in its interface. For each such element, the developer can select the left or the right interface –

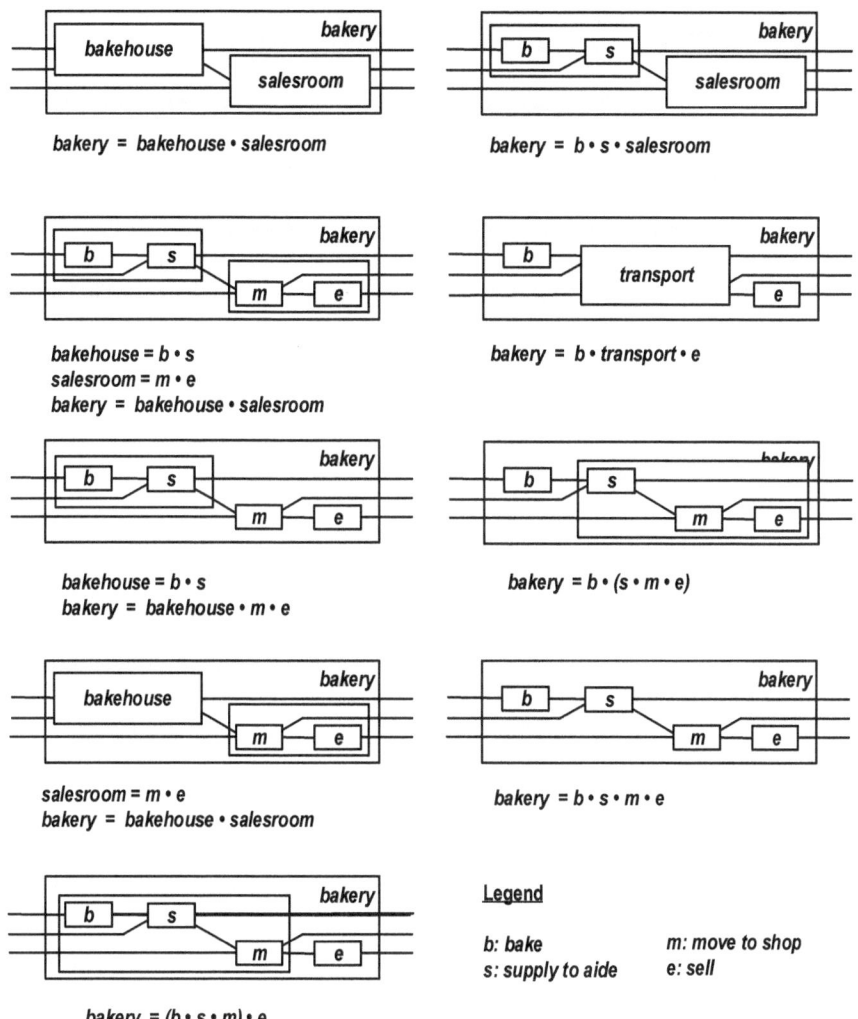

Fig. 12.2 Views of the bakery.

or, as we will see later, both interfaces. Proper choice of the interface(s) depends on the system's architecture, i.e. the interaction of the module with other modules.

The interior of an abstract module is usually refined to a set of composed modules. Refinement can be iterated hierarchically. HERAKLIT allows single modules to be refined locally; a tree-like structure is not required; modules of different refinement levels may be composed. In this way, various *views* of a system can be achieved. Figure 12.2 sketches typical such views for the bakery system as in Figures 2.3 and 2.6, pages 15 and 17, in Chapter 2.

The refinement process terminates with modules on the behavioral level.

The real-world- and data-elements

A real-world system usually operates not only with abstract data, but also with real-world items. This includes articles such as *bread* and *shelves* as exemplified in this monograph, but also documents such as *invoices*, *sales agreements*, et cetera. In a digital world, many items are additionally digitally represented as data elements. HERAKLIT considers and carefully distinguishes real-world items and their digital representation, as well as operations on those items and representations. This is conceptually covered by structures, signatures, and schemata, as discussed in Part III. During the design of a system model, there is no specific order of designing structures, signatures, and schemata on one hand, and architectural aspects on the other.

The behavioral level

To describe behavior, Petri nets are turned into modules. We distinguish three levels of behavioral modules:

- A behavioral module describes *one* concrete run, as exemplified in Section 4.2-4.4, pages 43ff. The case of interfaces with only places is prevalent; but see Figure 4.7, page 48, in Section 4.4. A place represents the occurrence of a state s, i.e. reaching and abandoning s, where s is a proposition. A transition represents the occurrence of an event. In the early phases of constructing a system model, it is worthwhile to design some particularly important such runs.
- A behavioral module describes a *system*. Elementary system modules focus (distributed) control, as in Sections 5.1, page 53f., and 5.3, page 56f. A place of an elementary system module represents a proposition. *General system modules* integrate items and data, as e.g. Figure 8.12, page 93, in Section 8.3. A place represents a predicate, applying to a changing set of items and data. A transition together with the inscriptions on its adjacent edges represents a set of occurrences of an event. This is achieved in a parameterized form, using the well-known feature of variables in terms. Design of systems traditionally starts with behavioral models. HERAKLIT suggests starting with architecture and runs, as described above.
- A behavioral module describes a *system schema*, as exemplified e.g. by Figure 10.2, page 109, in Section 10.2. A place of a system schema represents a *predicate symbol* of the underlying signature. Frequently, a system model is intended to represent just one system; nevertheless, a signature is assumed such that terms at edge inscriptions are a technically unambiguous concept.

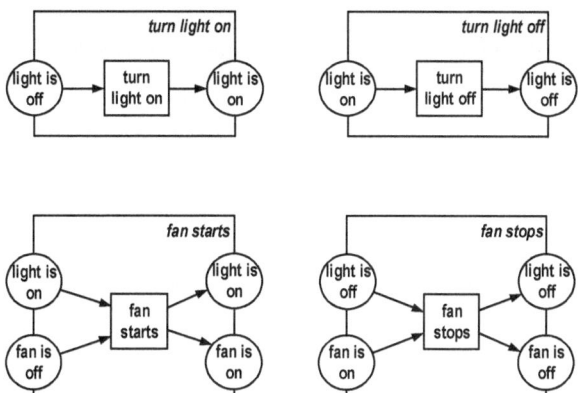

Fig. 12.3 The four steps of the light/fan system.

Example: from steps to a system module

To construct an elementary system module, it is often convenient to start by identifying typical propositions as places, and involved activities as transitions. Then, each activity is related to the involved propositions, thus generating a single step. These steps can be composed into typical runs of the intended elementary system module. Furthermore, to construct a corresponding elementary system module N, all occurrences of a proposition in all steps are merged, generating a place of N. The transitions and arcs of N are inherited from the steps.

As an example of this procedure, we start with a colloquial description of a system containing a *fan* and a *light*, as they are standard in modern bathrooms: In the case of the *fan is off*, when you turn on the light, after some time, the fan will start running. In this situation, if you turn off the light, the fan continues running for some time. Hence, in the case of the *fan is off*, when you turn the light on and off quickly, the fan will not start running at all. And in the case of the *fan on*, when you turn the light off and on the quickly, the fan will run continuously.

To model this system, we first extract the involved propositions from the description: *fan is off*, *fan is on*, *light is off*, and *light is on*. Furthermore, we identify four activities: *turn light on*, *turn light off*, *fan starts running*, and *fan stops running*. The effect of each activity on the propositions can be represented as a step, shown in Figure 12.3. Starting in a situation with both light and fan off, the system can evolve in different runs; it is non-deterministic. Figure 12.4 shows two such runs.

The system itself permits infinitely many runs. It is now easy to derive an elementary system module N, such that the runs of N are exactly the runs of the system: for each proposition p in the steps of Figure 12.3 merge all occurrences of p into one place of N. This results in the elementary system module of Figure 12.5. As an initially assumed state one may choose both the light and the fan off.

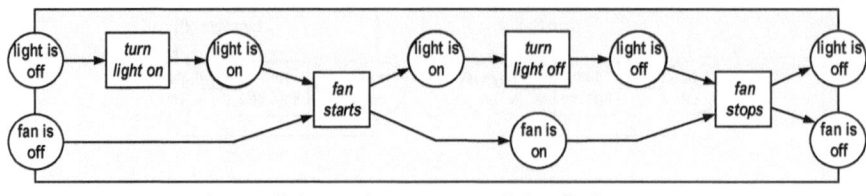

(a) turn light on • fan starts • turn light off • fan stops

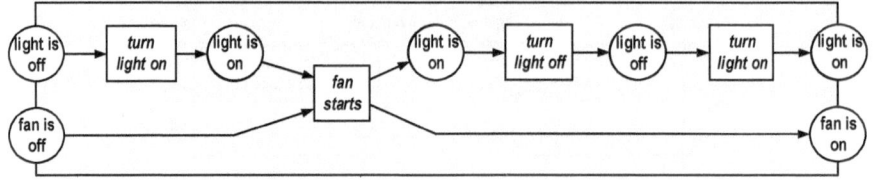

(b) turn light on • fan starts • turn light off • turn light on

Fig. 12.4 Two runs of the light/fan system.

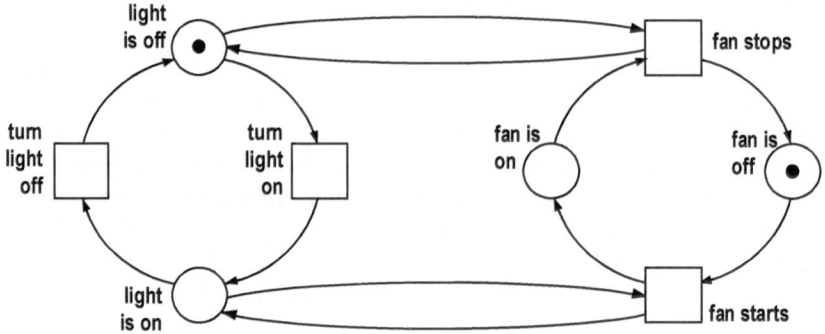

Fig. 12.5 The bathroom module.

Decomposition and design of large modules

In general, there are different ways to decompose a large system into reasonably designed modules. Different decompositions emphasize different aspects, symmetries, and asymmetries. We show this by the example of a model of a car dealer: After agreeing with a client on a car sale, the dealer initiates two processes: A process to organize credit for the client, and a process to prepare delivery of the car. After the bank agrees the credit, the car is delivered to the client. In the rare case of the bank's refusal of credit, the prepared car is returned to the warehouse. We show two different representations of the same elementary system module, composed from different sub-modules.

Figure 12.6 shows two modules:

- The car dealer's administration: the upper straight line shows a successful sale, from agreement with the client to archiving of success. As an exception, the

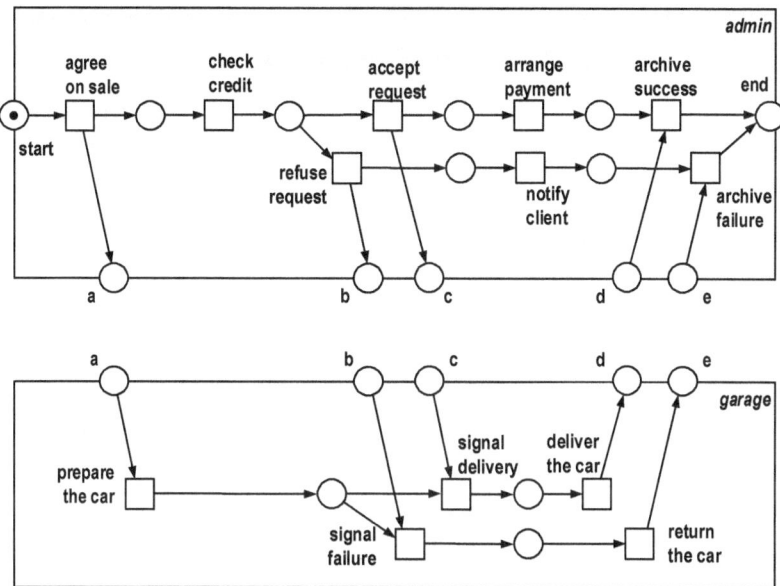

Fig. 12.6 Car sales composed from the modules *admin* and *garage*.

request for credit may be refused, as shown in the lower line. It is pictorially worthwhile to extend the right interface to the lower margin, hence the places with labels $a, \ldots, e$ of the module *admin* belong to the right interface *admin**.

- The garage, with cars to be sold: First, the car is prepared. Then, depending on the admin's signal, the car is either delivered or returned. The places in the module's upper margin belong to the module's left interface **garage*.

Figure 12.7 likewise shows two modules:

- the happy path, where a sale is agreed and successfully executed; the transitions on the lower margin belong to the module's right interface;
- the failure path, showing the effect of a refused credit request; the transitions on the upper margin belong to the module's right interface.

Both approaches result in the same behavioral module:

$$admin \bullet garage = happy\ path \bullet failure\ path . \qquad (12.1)$$

12.2 Graphical conventions

HERAKLIT models are graphs, with inscribed vertices and edges. A graph may be depicted in different ways. The chosen layout decisively affects the model's intuitive comprehensibility.

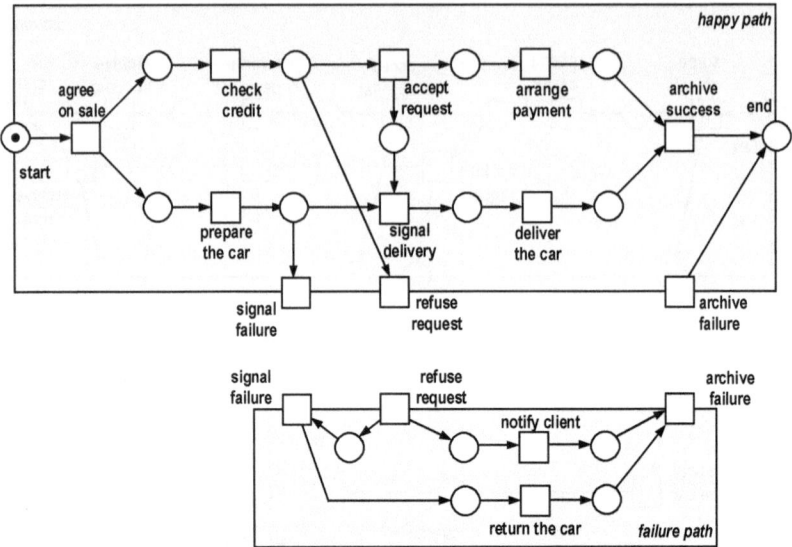

Fig. 12.7 Car sales composed from the modules *happy path* and *failure path*.

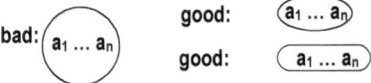

Fig. 12.8 Rounded vertices adjust to long inscriptions.

Compared to textual representations, diagrams offer a multitude of options. A modeler can and should utilize this freedom. At the same time, space is rare in diagrams. Here we suggest a bunch of recommendations to depict HERAKLIT modules as intuitively attractive diagrams. In a complex diagram, not all recommendations can be fulfilled simultaneously. So, compromises are inevitable. All diagrams in this text follow the following recommendations.

Vertices

As usual, Petri net places and transitions are depicted as circles and squares. Long inscriptions frequently conflict with circles; in this case, ellipses or ovals are favored, as in Figure 12.8. Likewise, a square may be replaced by a rectangle.

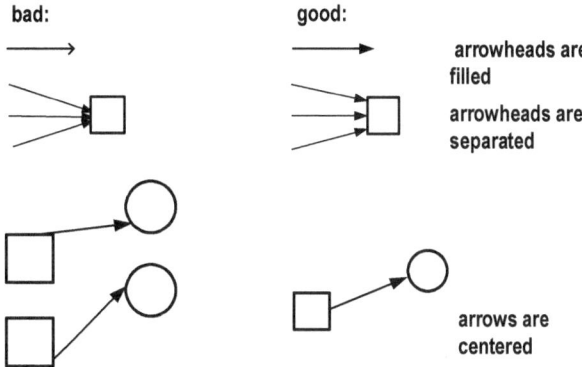

Fig. 12.9 Economy of arrows.

Arrowheads

For the sake of visibility and economy of space, an arrow head is dark and narrow, as in Figure 12.9 (top right). Usually, each vertex has a (not visible) center. Arrowheads must not overlap as in Figure 12.9 (top left). An arrow starts and ends virtually at the center of its adjacent vertex, as Figure 12.9 (right) shows. This principle can be ignored in favor of graphical clarity.

Curved arrows

Curved arrows can express symmetries, repetition, and other regular structures. A typical example is a cyclic structure with many places and transitions on its surface. Such structures should be depicted as (virtual) cycles or ellipses. Figure 12.10 shows examples.

Smallest elements

Each graphical representation has smallest elements that still must be clearly distinguishable. Sometimes, the smallest elements are indices, but also arrowheads and other components can be smallest elements. The difference in size between smallest and biggest elements should be kept reasonable. Figure 12.11 shows examples.

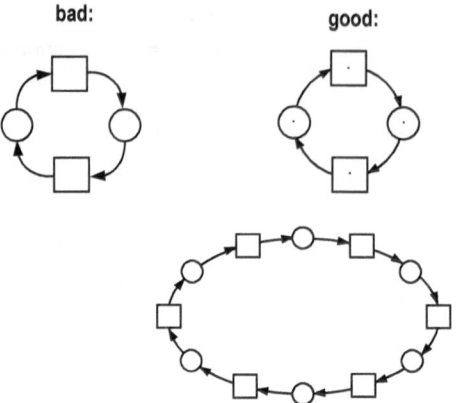

Fig. 12.10 Curved arrows on a virtual circle or ellipse.

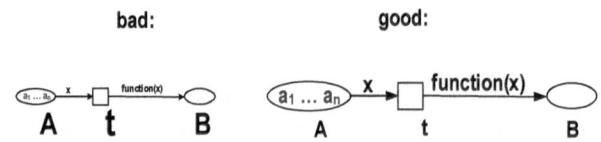

Fig. 12.11 Relative size difference must be small.

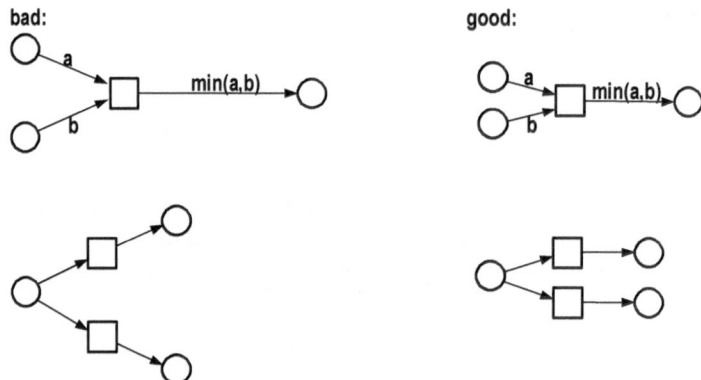

Fig. 12.12 No waste of space.

Space

A diagram should not waste space. Symbol fonts should be tight and sans-serif; we recommend *Arial Narrow Bold*. No edge should be unnecessarily long; blank areas should not be unnecessarily wide. Figure 12.12 gives examples.

Structure of diagrams

The most important flow of items, data, or control moves left-right, motivated by the conventional direction of reading. Frequently, one chooses up-down. This alternative depends on the size of the diagram: The width of a page is fixed; a well-proportioned diagram just fills this space. A slim diagram wastes space. A cyclic behavior may be drawn on a cycle or an ellipse. Exceptions, failure paths, et cetera may be irregularly arranged. A shaded background of a module's area is optically attractive. Areas and sub-areas alternate white and shaded backgrounds, such as e.g. in Figure 3.12, page 31, in Section 3.6.

Structure of diagrams

Related work to Part V

A myriad of recommendations for the early stages of software projects have been proposed [56]. In this context, the term *software model* usually refers to the process of constructing software, typically in the shape of a life cycle. This differs fundamentally from the construction of models.

Recommendations for the process of constructing models are less frequent, but emerging. A typical example is the blog [85] from Carnegie Mellon University, introducing *model-based systems engineering*, and suggesting a system construction process that incorporates the semi-formal *System Modeling Language* (*SysML*) as well as the UML. As [85] states:

"The model should demonstrate, in an easily comprehensible way, how the system must be built to be successful. Visualization is a key way to ensure comprehensibility. Visualizing abstract ideas enables people to take the leap of imagination that is needed to 'see' the system."

This also describes one of the goals of HERAKLIT. In addition, a HERAKLIT model is formally based, such that verification and automatic implementation of parts of the model are eventually feasible.

Closer to HERAKLIT are the patterns of [80]. Each of these patterns can be conceived as a HERAKLIT module in a very natural way. A lot of further approaches can be mentioned and discussed here. But to our best knowledge, compared to HERAKLIT, none of them is likewise comprehensive, visualized, and formally based.

Work on maturity models in software engineering began in the 1980s with [39]. Since then, a plethora of maturity models have emerged. Our proposal for digital transformation with HERAKLIT stands in this tradition. In the future, we would like to see testimonials, experience reports, and case studies using HERAKLIT for the digital transformation of the world. Established methods for developing a maturity model [7] could inform the development of the proposed maturity model.

In the conceptual modeling literature, numerous modeling guidelines have been elaborated. The so-called *Guidelines for Modeling* (GOM) developed by *Becker, Rosemann, Schütte*, and others, e.g. [8], is a particular ambitious and well-recognized framework. Their proposal consists of six general guidelines, e.g. principle of correctness, principle of relevance, and economic principle. The *GOM* are further specialized for different modeling framework, e.g. ERM and EPC. The hints and conventions for modeling given in Chapters 12.1 and 12.2 can be understood as one concrete instantiation of the GOM for HERAKLIT.

Furthermore, several aspects of modeling are studied from a behavioral science paradigm, e.g. the acceptance, diffusion, and understandability of modeling [38, 50, 55, 60]. This kind of research often confirms intuitive ideas that are also followed by the aforementioned conventions for HERAKLIT, e.g. an obscure and complicated diagram layout reduces the comprehensibility of a conceptual model. However, universal theoretical foundations for modeling from a behavioral science paradigm have yet to be developed. We would like to see HERAKLIT included in such future studies.

Conclusion and preview of forthcoming contributions

The reader of this book has got the background to grasp any kind of HERAKLIT models, in particular the case studies on the HERAKLIT homepage. In addition, Part V suggests rules of thumb to systematically construct HERAKLIT models.

Forthcoming texts will shed light on particularly useful aspects of modules, and on sets of modules. The expressive power of module composition is increased. The particularly important case of behavioral modules, both on the level of single behaviors and on the level of systems, will be studied in greater detail. A fundamental aspect of behavioral modules is the representation of important properties, their analysis, and their verification.

Application of HERAKLIT will be supported by:

- design patterns, including well-known patterns as in [80], but also patterns that exploit and refer to particular HERAKLIT constructs,
- distinguished case studies, as examples to be emulated,
- domain-specific HERAKLIT dialects for special classes of applications,
- software tools for the design, the analysis, and the verification of HERAKLIT.

P. Fettke, W. Reisig, *Understanding the Digital World*,
https://doi.org/10.1007/978-3-031-61898-7

References

1. G. Agha. Actors: A model of concurrent computation in distributed systems. *MIT Press*, 1986.
2. S. Alter. Work system theory: Overview of core concepts, extensions, and challenges for the future. *Journal of the Association for Information Systems*, pages 72–121, 2013.
3. F. Arbab. Reo: A channel-based coordination model for component composition. *Mathematical Structures in Computer Science*, 14(3):329–366, 2004.
4. J. C. M. Baeten, B. Luttik, and P. Van Tilburg. Reactive Turing machines. *Information and Computation*, 231:143–166, 2013.
5. P. Baldan, A. Corradini, H. Ehrig, and B. König. Open petri nets: Non-deterministic processes and compositionality. In H. Ehrig, R. Heckel, G. Rozenberg, and G. Taentzer, editors, *Graph Transformations, 4th International Conference, ICGT 2008, Leicester, United Kingdom, September 7-13, 2008. Proceedings*, pages 257–273. Springer, 2008.
6. J. C. M. Beaton and W. Weijiland. *Process Algebra*. Press Syndicate of the University of Cambridge, 1990.
7. J. Becker, R. Knackstedt, and J. Pöppelbuß. Developing maturity models for IT management. *Business & Information Systems Engineering*, 1(3):213–222, 2009.
8. J. Becker, M. Rosemann, and C. Von Uthmann. Guidelines of business process modeling. In *Business Process Management: Models, Techniques, and Empirical Studies*, pages 30–49. Springer, 2000.
9. L. Bernardinello, I. Lomazova, R. Nesterov, and L. Pomello. Soundness-preserving composition of synchronously and asynchronously interacting workflow net components. *Journal of Parallel and Distributed Computing*, 179:213–222, 2023.
10. E. Best, R. Devillers, and M. Koutny. *Petri Net Algebra*. Springer Science & Business Media, 2001.
11. E. Best and C. C. Fernández. Partially ordered sets. In *Nonsequential Processes: A Petri Net View*, pages 7–53. Springer, 1988.
12. D. Bjørner. Domain science & engineering - A foundation for software development (2023). *Revised edition of [17]. xii*, 2023.
13. U. Boker and N. Dershowitz. The Church-Turing thesis over arbitrary domains. In *Pillars of computer science: essays dedicated to Boris (Boaz) Trakhtenbrot on the occasion of his 85th birthday*, pages 199–229. Springer, 2008.
14. N. Bourbaki. The architecture of mathematics. *The American Mathematical Monthly*, 57(4):221–232, 1950.
15. M. Broy. A logical basis for component-oriented software and systems engineering. *The Computer Journal*, 53(10):1758–1782, 2010.
16. M. Bunge. *Causality and Modern Science*. Routledge, 4 edition, 2009.
17. P. P.-S. Chen. The entity-relationship model—toward a unified view of data. *ACM Transactions on Database Systems (TODS)*, 1(1):9–36, 1976.

P. Fettke, W. Reisig, *Understanding the Digital World*,
https://doi.org/10.1007/978-3-031-61898-7

18. S. Christensen and L. Petrucci. Towards a modular analysis of coloured Petri nets. In *International Conference on Application and Theory of Petri Nets*, pages 113–133. Springer, 1993.

19. L. de Alfaro and T. A. Henzinger. Interface automata. *SIGSOFT Softw. Eng. Notes*, 26(5):109–120, sep 2001.

20. H.-D. Ebbinghaus, J. Flum, W. Thomas, and A. S. Ferebee. *Mathematical Logic*. Springer, 1994.

21. S. Fendrich and G. Lüttgen. A generalised theory of interface automata, component compatibility and error. *Acta Informatica*, 56:287–319, 2019.

22. O. K. Ferstl and E. J. Sinz. Modeling of business systems using som. In P. Bernus, K. Mertins, and G. Schmidt, editors, *Handbook on Architectures of Information Systems*, pages 347–367. Springer, 2006.

23. P. Fettke and W. Reisig. HERAKLIT homepage. heraklit.dfki.de [Accessed: 2024-02-06].

24. P. Fettke and W. Reisig. Modelling service-oriented systems and cloud services with HERAKLIT. In C. Zirpins, I. Paraskakis, V. Andrikopoulos, N. Kratzke, C. Pahl, N. E. Ioini, A. S. Andreou, G. Feuerlicht, W. Lamersdorf, G. Ortiz, W. van den Heuvel, J. Soldani, M. Villari, G. Casale, and P. Plebani, editors, *Advances in Service-Oriented and Cloud Computing - International Workshops of ESOCC 2020, Heraklion, Crete, Greece, September 28-30, 2020, Revised Selected Papers*, pages 77–89. Springer, 2020.

25. P. Fettke and W. Reisig. Breathing life into models: The next generation of enterprise modeling. In H. Fill, M. van Sinderen, and L. A. Maciaszek, editors, *Proceedings of the 17th International Conference on Software Technologies, ICSOFT 2022, Lisbon, Portugal, July 11-13, 2022*, pages 7–14. SCITEPRESS, 2022.

26. P. Fettke and W. Reisig. Systems mining with heraklit: The next step. In C. D. Ciccio, R. M. Dijkman, A. del-Río-Ortega, and S. Rinderle-Ma, editors, *Business Process Management Forum - BPM 2022 Forum, Münster, Germany, September 11-16, 2022, Proceedings*, pages 89–104. Springer, 2022.

27. P. Fettke and W. Reisig. A causal, time-independent synchronization pattern for collective adaptive systems. *International Journal on Software Tools for Technology Transfer*, 25(5):659–673, 2023.

28. J. W. Forrester. *Industrial Dynamics*. MIT Press, 9th edition, 1977.

29. U. Frank. Multi-perspective enterprise modeling: Foundational concepts, prospects and future research challenges. *Software & Systems Modeling*, 13:941–962, 2014.

30. U. Frank, S. Strecker, P. Fettke, J. Vom Brocke, J. Becker, and E. Sinz. The research field "modeling business information systems": Current challenges and elements of a future research agenda. *Business & Information Systems Engineering*, 6:39–43, 2014.

31. H. J. Genrich and K. Lautenbach. System modelling with high-level Petri nets. *Theoretical computer science*, 13(1):109–135, 1981.

32. G. Grätzer. *Universal Algebra*. Springer New York, NY, 1968.

33. J. F. Groote and M. R. Mousavi. *Modeling and Analysis of Communicating Systems*. MIT press, 2014.

34. Y. Gurevich. Sequential abstract-state machines capture sequential algorithms. *ACM Transactions on Computational Logic (TOCL)*, 1(1):77–111, 2000.

35. D. Harel. Statecharts: A visual formalism for complex systems. *Science of computer programming*, 8(3):231–274, 1987.

36. D. Harel and R. Marelly. *Come, Let's Play: Scenario-Based Programming Using Lscs and the Play-Engine*. Springer, 2003.

37. C. A. R. Hoare. *Communicating Sequential Processes*. Prentice-hall Englewood Cliffs, 1985.

38. C. Houy, P. Fettke, and P. Loos. Understanding understandability of conceptual models–what are we actually talking about? In P. Atzeni, D. Cheung, and S. Ram, editors, *Conceptual Modeling. ER 2012*, pages 64–77. Springer, 2012.

39. W. S. Humphrey. Characterizing the software process: a maturity framework. *IEEE software*, 5(2):73–79, 1988.

40. I. Jacobson, G. Booch, and J. Rumbaugh. *The Unified Modeling Language Reference Manual*. Addison Wesley Longman, Inc., 1999.

41. I. Jacobson, G. Booch, and J. Rumbaugh. *The Unified Software Development Process*. Addison-Wesley, 1999.
42. P. Janich. *Handwerk und Mundwerk: Über das Herstellen von Wissen*. CH Beck, 2015.
43. K. Jensen and L. M. Kristensen. *Coloured Petri Nets*. Springer-Verlag Berlin Heidelberg, 2009.
44. S. Junginger, H. Kühn, R. Strobl, and D. Karagiannis. Ein Geschäftsprozessmanagement-Werkzeug der nächsten Generation - Adonis: Konzeption und Anwendungen. *Wirtschaftsinformatik*, 42(5):392–401, 2000.
45. G. Keller, M. Nüttgens, and A.-W. Scheer. Semantische Prozeßmodellierung auf der Grundlage "Ereignisgesteuerter Prozeßketten (EPK)". *Institut für Wirtschaftsinformatik (IWi), Universität des Saarlandes, Saarbrücken*, 89, 1992.
46. E. Kindler. A compositional partial order semantics for Petri net components. In *Application and Theory of Petri Nets 1997: 18th International Conference, ICATPN'97 Toulouse, France, June 23–27, 1997 Proceedings 18*, pages 235–252. Springer, 1997.
47. E. Kindler. On the semantics of EPCs: Resolving the vicious circle. *Data & Knowledge Engineering*, 56(1):23–40, 2006.
48. E. Kindler and L. Petrucci. Towards a standard for modular Petri nets: a formalisation. In *Applications and Theory of Petri Nets: 30th International Conference, PETRI NETS 2009, Paris, France, June 22-26, 2009. Proceedings 30*, pages 43–62. Springer, 2009.
49. R. B. Kline. *Principles and Practice of Structural Equation Modeling*. Guilford Press, 5 edition, 2023.
50. J. Krogstie, G. Sindre, and O. I. Lindland. 20 years of quality of models. In *Seminal Contributions to Information Systems Engineering: 25 Years of CAiSE*, pages 103–107. Springer, 2013.
51. D. A. Lamb. Idl: Sharing intermediate representations. *ACM Transactions on Programming Languages and Systems (TOPLAS)*, 9(3):297–318, 1987.
52. L. Lamport. Who builds a house without drawing blueprints? *Commun. ACM*, 58(4):38–41, 2015.
53. A. Martin, M. Magnaudet, and S. Conversy. Computers as interactive machines: Can we build an explanatory abstraction? *Minds and Machines*, 33(1):83–112, 2023.
54. J. Mendling. *Metrics for Process Models: Empirical Foundations of Verification, Error Prediction, and Guidelines for Correctness*, volume 6. Springer Science & Business Media, 2008.
55. J. Mendling, H. A. Reijers, and W. M. van der Aalst. Seven process modeling guidelines (7pmg). *Information and software technology*, 52(2):127–136, 2010.
56. J. Michael, D. Bork, M. Wimmer, and H. C. Mayr. Quo vadis modeling? *Softw. Syst. Model.*, 23(1):7–28, 2024.
57. R. Milner. Calculi for interaction. *Acta informatica*, 33:707–737, 1996.
58. R. Milner. *Communicating and Mobile Systems: the Pi Calculus*. Cambridge university press, 1999.
59. R. Milner. Bigraphs and their algebra. *Electronic Notes in Theoretical Computer Science*, 209:5–19, 2008.
60. D. Moody. The "physics" of notations: Toward a scientific basis for constructing visual notations in software engineering. *IEEE Transactions on software engineering*, 35(6):756–779, 2009.
61. H. Mrech, E. Ortner, U. Raape, S. Overhage, S. Sahm, A. Schmietendorf, T. Teschke, and K. Turowski. Standardized specification of business components. *Component Oriented Business Application Systems*, 2002.
62. P. Naur and B. Randell. Software engineering: Report of a conference sponsored by the NATO Science Committee, Garmisch, Germany, 7th-11th October 1968. Technical report, Newcastle University, 1968.
63. OMG. Business Process Model and Notation (BPMN) version 2.0.2, 2014.
64. H. Österle. *Business in the Information Age: Heading for New Processes*. Springer, 1995.
65. D. L. Parnas. A technique for software module specification with examples. *Communications of the ACM*, 15(5):330–336, 1972.

66. S. Patig. Evolution of entity–relationship modelling. *Data & Knowledge Engineering*, 56(2):122–138, 2006.

67. J. Pearl. *Causality*. Cambridge University Press, 2nd edition, 2009.

68. J. L. Peterson. *Petri Net Theory and the Modeling of System*. independently published, 3rd edition, 2019.

69. C. A. Petri. *Kommunikation mit Automaten*. PhD thesis, Universität Bonn, 1962.

70. C.-A. Petri. Non-sequential processes, interner bericht isf-77-5. *Gesellschaft für Mathematik und Datenverarbeitung*, 1977.

71. Petri Nets World. www.informatik.uni-hamburg.de/TGI/PetriNets/ [Accessed: 2024-02-06], 2024.

72. C. J. Petrie. *Web Service Composition*. Springer, 2016.

73. J. Rathke, P. Sobociński, and O. Stephens. Compositional reachability in Petri nets. In *Reachability Problems: 8th International Workshop, RP 2014, Oxford, UK, September 22-24, 2014. Proceedings 8*, pages 230–243. Springer, 2014.

74. J. C. Recker, R. Lukyanenko, M. Jabbari Sabegh, B. Samuel, and A. Castellanos. From representation to mediation: a new agenda for conceptual modeling research in a digital world. *MIS Quarterly: Management Information Systems*, 45(1):269–300, 2021.

75. W. Reisig. On the expressive power of Petri net schemata. In *International Conference on Application and Theory of Petri Nets*, pages 349–364. Springer, 2005.

76. W. Reisig. *Understanding Petri Nets*. Springer, 2013.

77. W. Reisig. Associative composition of components with double-sided interfaces. *Acta Informatica*, 56(3):229–253, 2019.

78. D. Ritter, S. Rinderle-Ma, M. Montali, and A. Rivkin. Formal foundations for responsible application integration. *Information Systems*, 101:1–24, 2021.

79. A. W. Roscoe. *Understanding Concurrent Systems*. Springer Science & Business Media, 2010.

80. N. Russell, W. M. Van Der Aalst, and A. H. Ter Hofstede. *Workflow Patterns: the Definitive Guide*. MIT Press, 2016.

81. K. Sandkuhl, H.-G. Fill, S. Hoppenbrouwers, J. Krogstie, F. Matthes, A. Opdahl, G. Schwabe, Ö. Uludag, and R. Winter. From expert discipline to common practice: a vision and research agenda for extending the reach of enterprise modeling. *Business & Information Systems Engineering*, 60:69–80, 2018.

82. D. Sannella and A. Tarlecki. *Foundations of Algebraic Specification and Formal Software Development*. Springer, 2012.

83. A.-W. Scheer. *ARIS-Business Process Frameworks*. Springer Science & Business Media, 3 edition, 2012.

84. A.-W. Scheer. *The Composable Enterprise: Agile, Flexible, Innovative: A Gamechanger for Organisations, Digitisation and Business Software*. Springer, 4th edition, 2023.

85. N. Shevchenko. An introduction to model-based systems engineering (mbse). *Software Engineering Institute Blog*, 2020.

86. P. Sobociński. Representations of Petri net interactions. In *International Conference on Concurrency Theory*, pages 554–568. Springer, 2010.

87. J. Sterman. *System Dynamics: Systems Thinking and Modeling for a Complex World*. MIT Press., 2000.

88. V. C. Storey, R. Lukyanenko, and A. Castellanos. Conceptual modeling: Topics, themes, and technology trends. *ACM Journal on Computing and Cultural Heritage*, 2023.

89. P. Suppes. *Introduction to Logic*. Dover Publications, Mineola, N.Y., 1957.

90. P. Suppes. *Representation and Invariance of Scientific Structures*. CSLI Publications, 2002.

91. T. Teufel and G. Keller. *SAP R/3 Process Oriented Implementation: Iterative Process Prototyping*. Addison-Wesley, 1998.

92. W. M. P. van der Aalst. Formalization and verification of event-driven process chains. *Information and Software Technology*, Volume 41, Issue 10:639–650, 1999.

93. Y. Wand and R. Weber. Research commentary: Information systems and conceptual modeling – a research agenda. *Information Systems Research*, 13(4):363–376, 2002.

94. P. Wegner. Why interaction is more powerful than algorithms. *Communications of the ACM*, 40(5):80–91, 1997.

95. M. Weske. *Business Process Management: Concepts, Languages, Architectures.* Springer Berlin, Heidelberg, 3rd edition, 2019.
96. N. Wiener. *Cybernetics.* Martino Fine Books, 2nd edition, 2013.
97. R. Winter. Working for e-business – the business engineering approach. *International Journal of Business Studies*, 9(1):101–117, 2001.

References

196

Index

© The Author(s), under exclusive license to Springer Nature Switzerland AG 2024
P. Fettke, W. Reisig, *Understanding the Digital World*,
https://doi.org/10.1007/978-3-031-61898-7